YO-BYF-855

Cost-Effective Patenting™
The Essential Business Guide To
Patenting Strategies

By Robert D. Fish, Esq.

Toothbrush diagrams prepared by Bryan Hersey

Published by
SkyMan™ Publications

Second Edition

Printed in the United States of America
ISBN 0-9719165-0-0

TABLE OF CONTENTS

COST-EFFECTIVE PATENTING
BY ROBERT D. FISH

Introduction

The importance of intellectual property (IP) has increased tremendously over the last few decades. In many companies the patents, trademarks, copyrights, and trade secrets rights comprise a substantial proportion of the entire value of the company.

Unfortunately, the cost of protecting intellectual property has also risen dramatically. In start-up companies the IP expense can easily grow to such an extent that it adversely affects research and development (R&D) and other critical functions. When someone does finally pull the plug on IP expenses, the process is often so far along that valuable intellectual property gets flushed down the drain. Applications that have already been filed are abandoned, and applications for new inventions that deserve to be protected never get filed at all.

Such situations can be largely avoided by employing *cost-effective patenting strategies*. The application of cost-effectiveness may be relatively new to the world of intellectual property, but the basic concept is standard fare in the rest of the business world. Any good business person focuses on cost-effectiveness in renting office space, in hiring employees, and in purchasing a photocopier. Why should the very same concept be forgotten as soon as one turns to his lawyer? The answer seems to be that intellectual property in general, and patenting in particular, are seen as "magical arts" that fall outside the understanding of ordinary business and technical people. Well, there *is* a great deal of "art" to these things, but there is certainly no magic. Both management and inventors have a duty to understand and control the process.

This book provides a synopsis of numerous cost-effective patenting strategies. Some of these strategies are cost-effective because they produce *direct cost savings*. Other strategies are cost-effective because they produce *stronger patents* and provide *better market position*.

Chapter I begins by discussing why patenting is so expensive. It turns out that many of the costs are inherent in the system, and cannot be realistically reduced. But many other costs can be reduced, and such reductions can be very significant. It is not at all unusual for the strategies contained herein to save a client with even a modest portfolio hundreds of thousands of dollars.

Chapter II reviews the earliest strategy decisions involved in patenting. Chief among the early concerns is how much time and effort should go into searching for prior art (previous disclosures or uses of the invention). There are many considerations here, including the closeness and aggressiveness of the competition, the extent of R&D efforts that one expects to allocate to the new technology over the next year or two, the importance of having an early search report to secure funding, and the speed with which the company wants to get a product on the market. Depending on the situation, the better decision might be to forgo a pre-filing patentability search, or to do just the opposite and run an extensive patentability search. Similarly, it might be better to invest in a non-infringement opinion, or to completely skip such an opinion.

Chapter III goes on to describe the next major strategy decisions, which involve what type of application to draft, and how it should be drafted. Yes, different attorneys have different styles, and indeed the styles used for *patent applications should vary* according to the technology, the closeness of the prior art, and several other factors. But there are guidelines that apply to substantially all circumstances.

One mark of a good application is that it is short and to the point. Anyone can write something that is long and complicated. The more difficult and more valuable task is to make it short and simple.

Another hallmark of a well-written patent application is that it works as a sales presentation. As discussed in Chapter III, the background section of a patent application should convince the patent examiner that the invention addresses a long-standing, difficult problem in the field. The detailed description should then go on to clearly define the inventive features, rather than get bogged down with excessive technical detail.

Of course, the claims are the most important aspect of a patent. Chapter IV explains how to distinguish between claims that are crafted with the clear purpose of securing a valuable space in the marketplace, and claims that merely focus on technical aspects of the invention. The "market-centered" approach is readily seen to be superior to the usual "invention-centered" approach.

Chapters V and VI focus the reader's attention on the importance of having short claims, and of utilizing only a small number of independent claims. Quantity is a very poor proper substitute for quality in patent claiming.

Chapter VII addresses the importance of claiming with litigation in mind. Patents are not self-enforcing, and valuable patents are often subject to litigation. It is extremely important to write the claims so that they are likely to withstand scrutiny at trial, both from a technical standpoint and from the standpoint of being understandable by non-technical judges and juries.

Chapter VIII addresses foreign filing. While this topic may seem to be a distant concern at the onset of the patenting process, it can very quickly become a major issue, easily increasing the patenting cost by ten or twenty fold. Knowing where and what to file early in the game requires the use of filing strategies that provide a rapid determination of patentability, such as the PCT first filling strategy, and aggressive use of petitions to make special. Failure to use those strategies means that the foreign filing becomes due even before the applicant receives the first office action, and has any feedback on what claims are likely to be allowed. A good foreign filing strategy also focuses on writing claims in the US or PCT applications that are readily transportable into the foreign countries. Many foreign patent offices, for example, won't allow more than a few independent claims. Prosecuting a great many independent claims in the US and PCT patent offices can waste a great deal of money in redrafting the claims for foreign filing.

Chapter IX addresses the inventive process itself. It turns out that technology often evolves in a predicable manner, and a patent attorney with an eye towards creativity can provide invaluable assistance in the patenting process.

Now it is true that some detractors will say that that cost-effectiveness should not be a primary focus of the patenting process, and that a company should spend whatever it takes to secure the best portfolio possible. That approach is misguided, and as one might expect comes mostly from the very patent attorneys who benefit the most from high patenting budgets. *The truth is that a cost-effective patenting strategy is superior to a wasteful strategy both in terms of money spent and in terms of results.* For one thing, forcing the patent attorney to be cost-effective requires him/her to distill the inventive subject matter down to its essence. That invariably leads to better patent claims. For another thing, a cost-effective patenting strategy goes a long way towards providing sufficient funding to file on all relevant inventions. It does little good to shortchange later developments because too much of the IP budget was spent on earlier applications.

The path is not always comfortable. Experience has shown that those who come to appreciate the strategies described in this book are often faced with the grim reality that they have already wasted a great deal of time, money and effort by filing previous patent applications in a manner that was not cost-effective. But that is all water under the bridge. The answer is to change course now, and to ensure that future work is performed cost-effectively .

Chapter I - Why Patent Expenses Are So High

It usually costs about $6,000 to $10,000 to get a single patent application on file, and another $5,000 to $10,000 to get the patent issued. Those expenses are only for the US. Foreign filing costs more ... a lot more! It typically costs another $100,000 or more to get your patent issued in the major foreign countries. Still further, those costs can be multiplied several fold as the original patent application spawns child and grandchild applications. The cost-effective approach is to minimize these costs by clever use of provisional applications, cost-effective claim drafting, and by carefully picking one's battles with the patent office.

When people first get started in the patent world they are often told that the cost of securing a patent is about $10,000 - $15,000. That's not too far off for many types of applications. As detailed in the first section below, the cost of getting a formal patent application on file is about $6,000 - $8,000, the costs of arguing (prosecuting) the application before the patent office is $1,500 to $3,000, and the cost of having the patent issue is yet another $750 (assuming small entity status). From start to finish, one often spends about $10,000 at the low end to secure a US utility patent, and up to $20,000 or more at the high end.

Cost Of Getting a Patent Issued

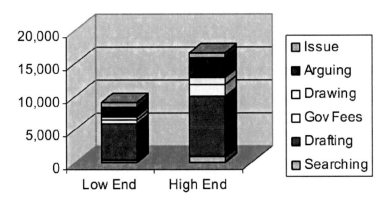

Figure 1

The problem is that such charges are often only the beginning. As described in the succeeding section, that single utility application may easily spawn numerous divisionals, which can increase the costs many fold. Add to that the costs of foreign prosecution, and what originated as a $10,000 budget very quickly grow to $750,000 or more.

A) Costs of Getting A Primary Application On File

The total cost of getting a formal patent application *on file* is usually about $6,000 - $10,000, depending on whether one is filing a utility or a PCT application. The breakdown is as follows:

Costs Of Filing A Patent Application

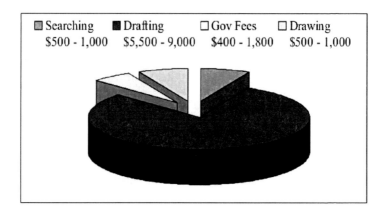

| ▣ Searching | ◼ Drafting | ☐ Gov Fees | ☐ Drawing |
| $500 - 1,000 | $5,500 - 9,000 | $400 - 1,800 | $500 - 1,000 |

Figure 2

- An experienced patent drafter takes two or four hours to search, and another ten to twelve hours to draft a typical patent application. At about $400 and hour that comes to about $6,000. Less experienced attorneys (or patent agents) may charge less per hour, but are typically slower, so that the total cost remains the same. Of course, drafting costs on very complicated inventions may run higher, as much as $10,000 or more.

Some firms charges as much as $30,000 to draft a typical patent application, but such high fees are usually unjustified, as demonstrated below.

- On top of the drafting charges one needs to add the filing costs. The government filing fee is about $380 for a utility application ($720 for so-called large entities that have more than 500 employees), and about $1,250 - $1,800 for a PCT (Patent Cooperation Treaty) application. The government filing fee for the utility application is much less than the PCT application because the entire process is handled by the USPTO (United States Patent and Trademark Office). PCT applications are handled by both the USPTO and WIPO (World Intellectual Property Organization) in Geneva, Switzerland.

- There are also drawing charges, which run about $150 per sheet. Most applications can be drafted to require only two or three sheets of drawing, so the total drawing charges should be less than about $500.

Of course, some applications are extremely difficult to draft, and then the costs go up. Usually the added difficulty arises from one of three circumstances. One very common cause for high drafting charges is that the invention is very simple. A new pair of scissors, for example, is incredibly difficult to patent broadly. The concept of scissors has been around for so long that the prior art is a virtual minefield of trouble spots.

Another common cause for high drafting charges is that the inventor keeps changing his mind about what is to be claimed. It is one thing to work with an inventor to derive claims, draft the application, and then get final sign-off with relatively few changes. It is quite another thing to spend the better part of a month discussing claims and writing the application, only to start the process all over again when the inventor decides that he wants to claim something different. We worked on an application recently where the costs doubled (at least) because the very process of ferreting out the invention caused everyone to realize that the invention was impractical as previously contemplated.

The third common cause for high patent application costs is that the inventor has a large number of designs and/or a huge amount of undigested data to consider. Running through all that information can be incredibly time consuming. Patent attorneys often take on the chore happily because they don't otherwise have a full plate. But the task is best done by the inventor, or at most by the inventor with help from the patent attorney or others.

B) <u>Costs of Getting A Secondary Application On File</u>

A first (i.e. primary) utility or PCT application is often followed by additional (i.e., secondary) applications, such as divisionals, continuations, CIPs, and so forth. The primary and secondary applications can quickly form a complicated *family* of applications. These other applications are usually much less expensive to file than the primary application.

Patent Families Can Be Complicated

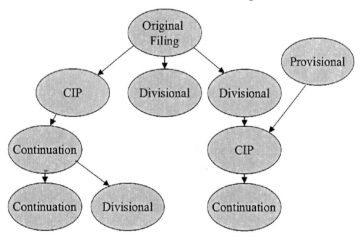

Figure 3

- One of the most common of these secondary type of applications is a "divisional". Divisionals are basically "child applications" based on a "parent" application, and necessarily have the same specification as the parent. The patent attorney cannot change the specification significantly even if he wants to, so all the effort goes into changing the claims. That sometimes costs another one or even two thousand dollars, but more commonly the patent attorney just files claims in the divisional that were withdrawn or rejected from the parent. Thus, the cost for filing a divisional application, including filing fees and paralegal charges, should usually be less than $1,000 (for a small entity).

- A "straight" continuation is akin to a divisional, and confusingly has gone by several different names over the years, including continuation, file wrapper continuation (FWC), and request for continuing application (RCE). These are merely re-filings or continued prosecution of the previous application, necessitated by repeated rejections of the pending claims by the patent office. Here again only the claims are amended, not the specification. The cost for filing a continuation application, including filing fees and paralegal charges, should usually be less than $1,000 (for a small entity).

- There is still another type of child application called a "continuation-in-part" or C-I-P for short. C-I-P applications use an existing application as a springboard, and then add additional subject matter to the specification. C-I-P applications also almost invariably add new claims. These applications are intermediate in difficulty (and cost) between a new utility or PCT application, and a divisional or straight continuation.

- A provisional patent application is an informal application. It never issues as a patent, and indeed is never even examined. Provisionals are basically placeholders. Their purpose is to give the applicant a filing date to which he can later claim priority in a formal application. Unless a formal application (utility or PCT) is filed within a year of the provisional filing date, the provisional gets archived in some warehouse at the one-year anniversary. Provisionals can be as simple as a single paragraph, or as complete as a formal application (more on how to use those choices cost-effectively in Chapter IV). Thus, the patent attorney charges for preparing a provisional vary widely. In some instances we have charged as little as $250, such as when we merely added a cover sheet to pages copied from a lab notebook. On the flip side we have also charged several thousands of dollars to draft a full utility application, which we nevertheless decided to file as a provisional. Provisionals currently have a filing fee of only about $75 for a provisional ($150 for large entities).

- A design application focuses on the ornamental (non-functional) appearance of something. Thus, designer sunglasses may often be protected with design patents (and possibly product configuration trade dress as well). There is nothing at all to writing the claims, because every design patent has only one claim, and that claim is always worded in substantially the same manner. The claim always reads along the lines of "A <something> as shown and described in the accompanying drawing". Indeed, the cost of a design application is not in the effort spent on the specification and claims, but almost entirely in the cost of preparing the many pages of drawing.

These should always be done by a professional patent drawing draftsman because of the many nuances peculiar to drawing design patent applications. Among other things one must be very careful to use as little detail as possible, and to depict non-claimed subject matter using dashed lines. Filing charges for a design patent, including drawing sheets, is usually under $1000.

C) <u>Variations In Charges From Firm to Firm</u>

There are firms that charge a lot more than the fees and costs identified above. Our office once assumed responsibility for a family of applications where the law firm had charged about $2000, in additional to government filing fees, to basically *photocopy* a US utility application and file it as a PCT. There weren't any substantive changes in the entire application! That is just absurd. Incidentally, before we got involved the law firm had spent years trying unsuccessfully to get claims allowed. I called the examiner, and in one day got broader claims allowed than had previously been rejected. Not only did the previous firm charge way too much money, but the previous patent attorneys never really understood where the novelty lay, which meant that they wasted many years and tens of thousands of dollars fighting windmills.

Some firms routinely charge $20,000 and even $30,000 to draft a patent application. Those charges are absurd. Just think about it. Let's say a patent attorney having a medium level of experience charges $300 per hour. If he/she drafts a $30,000 application, that means the attorney billed 100 hours to do the work. That's 2.5 solid weeks of work. Except in the most unusual circumstances, no competent patent attorney spends that much time writing a single application. A more realistic distribution of fees is shown below.

Distribution of Patent Application Charges

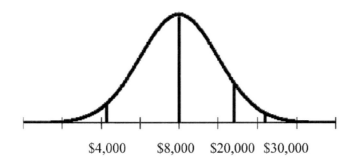

$4,000 $8,000 $20,000 $30,000

Figure 4

19

Of course, there are lots of tricks to justify the higher costs. Sometimes the trick is to file a very long application. Some firms routinely submit patent applications that are 80 pages long, and even 100 or pages or more at times. Yes, there are possible circumstances that justify such a long application, but it doesn't justify a high price. If the length of the application is due to an extensive listing of preferred embodiments or experiments (as is often the case in the pharmaceutical and other chemical fields), then the attorney can't justify the cost because those long listings are almost invariably compiled by the inventor(s), not the attorney. If the length is for any other reason, then it is still not justified because the length of the application is the result of the attorney's failure to figure out where the novelty lies. A complicated patent application is worth less, not more. Anyone can make something difficult. It takes hard work to make something simple.

Another common trick is to bill large amounts for work of a "supervising" attorney, who doesn't (or shouldn't) spend anywhere near the amount of time billed. One patent attorney told me that he regularly drafted applications for about $5,000 of time, and that the firm billed about $25,000 for his work! These things are just unconscionable.

D) <u>Costs of Arguing With The Patent Office</u>

Another area where people waste a lot of money is in prosecution (i.e., arguing back and forth with the patent examiner). In US prosecution an applicant is usually entitled to two bites at the apple; one response after a first substantive office action, and then a second response after a second substantive office action. Office actions for technicalities such as missing signatures in filing the application, or restriction requirements (see below) don't count towards the two bites. The second or third office action is almost always a final rejection, and prosecution can only continue by filing a continuation.

Fighting The Patent Office Is Not Always The Best Use of Funds

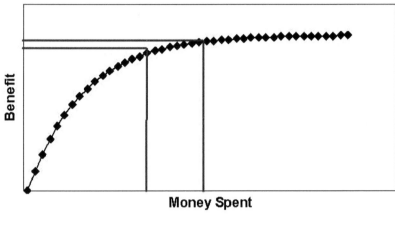

Figure 5

The arguing phase is usually completed within about two years from the first office action, and costs less than about $2500. But the process can go much longer. I once took over prosecution of a patent application that had been languishing in the patent office for ten years! The cost to the client in that case, including a series of continuations, was tens of thousands of dollars. I have also seen incredibly inefficient prosecution, which costs the client many thousands of dollars more than it should.

The key is to know when to fight and when not to fight. Attorneys are happy to carry the battle to the death, but of course it's your money that they are fighting with. There are diminishing returns here just as anywhere else.

The problem from a cost-effectiveness standpoint is often that patent attorneys view the patent examiners in a derogatory light. Even though the patent attorneys are ever so polite to the examiners, (as they must be under the patent office rules), they often fail to take the office actions seriously. That results in arguments that sound like two children arguing. One says "x", and the other says "not x". Then the first one repeats "x", and the other repeats "not x". This can go on for years, with the patent attorney developing a thick file to justify all his work. The cost, of course, is borne by the applicant, and runs many thousands of dollars. If your patent application is not issued after a first continuation, you have to seriously consider whether you have the right strategy.

The solution is to do the difficult mental work of figuring out where the invention lies relative to the prior art, and to write concise, commercially relevant short claims that focus on the point of novelty. The emphasis should not be on technical distinctions, but on claiming improvements that have significance in the marketplace (see Chapter V). Some of the tactics for doing so are outlined below, and additional details of these strategies, including working examples and tutorials, can be found in my upcoming companion book, provisionally entitled: "*The Art And Science Of Patent Prosecution*". Once the patent attorney clearly understands *and claims* the novelty over the prior art, the patent examiners are usually supportive and are perfectly willing to grant allowance.

It is true, of course, that high charges in some prosecutions are justified. From time to time an applicant pulls a bad examiner, and costs run up because the examiner doesn't know, or won't follow the law. I once had an examiner who insisted upon rejecting a dependent claim on the grounds of obviousness, while deeming the independent claim allowable! Such a rejection is completely illogical and wrong.

In other instances patent prosecution charges may validly run up because the patent attorney is trying to keep a continuing application alive, after all claims in the parent were issued. In that case the continuing application is an insurance policy, allowing new claims to be filed without loss of priority in the event a competitor figures out how to circumvent the issued claims. High prosecution costs can also be justified where the applicant chooses to keep his application pending rather than have narrow claims issue. In such instances the patent pending status can be more valuable in the marketplace than an issued patent with narrow claims.

E) One Application Can Easily Spawn Numerous Family Members

As we have seen, the cost of getting a US patent application on file should be $8,000 - $10,000. In rare cases the cost can go up to $15,000. By the time the patent issues, the total cost may run from $10,000 to as high as $20,000. So far so good.

A Single Application Is Reasonable

Figure 6

But many companies want to file foreign applications. As an example, it is not at all unusual for patent applications on a medical device, to be filed in Europe, Japan, China (PRC), Korea, Canada and Mexico. In that event securing the foreign patents raises the cost by another $10,000 per country. It is not at all unusual that the $10,000 - $20,000 budgeted for a US patent will spawn $100,000 to $150,000 in foreign charges.

Cost Can Still Be Reasonable With Foreign Filings

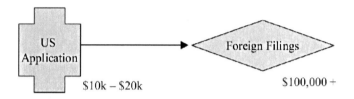

Figure 7

Now let's look at what happens when a single US application spawns both divisionals and foreign filings. Instead of about $100,000 in total charges, the foreign filings with multiple divisionals raise the cost to $350,000 or more.

Divisionals Raise The Cost Considerably

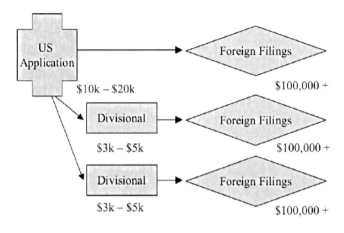

Figure 8

Finally, look at what happens when there are multiple US applications. The original $10,000 - $20,000 US patenting costs runs up a bill nearing a million dollars. Of course, for most individuals, and many companies, those numbers are just unacceptable. This is why people should be focusing on cost effectiveness. There are strategies for achieving the desired results, without triggering all that cost.

Multiple Applications Make It Much Worse

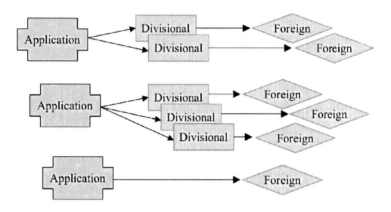

Figure 9

Chapter II - Search The Prior Art Cost-Effectively

Effective prior art searching is critical to filing a good patent application. On the other hand, not all applications should be searched, and not all searches should be particularly thorough. Cost-effective patenting requires conscious decisions as to when to search, what kind of search to perform, and how deep the search should go. Those decisions involve many different factors, from susceptibility of the field to keyword searching, to the impact of the search on research and development efforts. Searching should usually be iterative to avoid unnecessary effort. Although most outside searching services are not worth the money, there are strategies for securing very reliable searches, both from patentability and non-infringement (right-to-use) standpoints. Some of the best strategies involve having the PCT (International) or EPO (European) patent offices conduct your searches for you. Here is how to search cost-effectively.

A) Choose The Level Of Search That Is Best For You

What level of search should you do? The answer depends entirely upon your specific circumstances.

What Level Of Search Is Best?

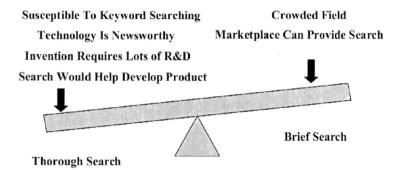

Susceptible To Keyword Searching

Technology Is Newsworthy

Invention Requires Lots of R&D

Search Would Help Develop Product

Crowded Field

Marketplace Can Provide Search

Brief Search

Thorough Search

Figure 10

- <u>Is the search readily susceptible to
keyword searching?</u> Searching usually
works well only where your invention
(i.e. the improvement over the prior
art) can be readily described with a
few keywords. As an example, a
clothing perspiration shield that sticks
to the skin would be readily
searchable, because one could search
for a combination of perspir*, shield
or cover, and adhere* or stick*.
Searching for suntan lotion packaged
in single use packets the way mustard
and ketchup are sold is also readily
searchable. One would just search for

a combination of the terms suntan, packet, and possible single-use.

- Is the field crowded? Patentability searching often works poorly in crowded fields, even if the invention is readily described by keywords. A new baby toy in which shaped pieces are inserted into correspondingly shaped holes is probably not worth searching. There are so many competitors, from well-known companies to local craftsmen, that anything more than a quick search is probably a waste of money.

- Would a thorough search help in developing the product? Sometimes it is useful to conduct a very thorough search even though the field is crowded and the subject matter is difficult to search. The benefit is not so much determining if the idea is patentable, but identifying what is known in the field to help guide the direction of R&D.

  The development of technology is like building an arch. No matter how many blocks are used to build the foundation of the arch, it won't stand up by itself until someone inserts the keystone at the top. In technology, there may be hundreds or even thousands of people working on a project, each adding to the foundation. But the person who adds that last missing piece gets the patent and all the credit, even though that keystone pieces is almost always quite small compared to the whole.

The One Who Adds The Keystone Gets All The Credit

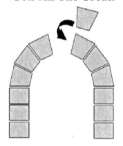

Figure 11

A good search can help place prior efforts into perspective, to help your group devise that final piece that makes the technology commercially viable.

- Is the technology likely or unlikely to be published? The various keyword databases are only useful in searching inventions that are published. For example, using keywords to search for the use of a particular sewing stitch on the hem of a swimsuit would be almost impossible. Swimsuits have been manufactured for well over a hundred years, with different manufactures regularly trying out new stitches without ever publishing information on what they are doing. Use of one stitch rather than another would therefore not likely be included in the local newspaper, in a patent, or in any publication for that matter.

- Does the invention require a lot of R&D? In some instances a company is planning to spend hundreds of thousands or even millions of dollars on R&D for a product. Knowing early on that the product is not patentable can save huge sums of money, or at least re-direct that money to something more worthwhile. In those instances the cost-effective approach is to invest in a thorough search as soon as possible, and update the search as the research develops.

- **Can the marketplace provide the search?** If the invention is fairly simple, and there is little or no additional conceptualization to be done, then the most cost-effective strategy is probably to file an inexpensive provisional application to secure patent pending status, and then go out in the marketplace and try to find a licensee or customer. If the product is old, those people will surely tell you, and you can abandon the application. If everyone loves the invention, then it is cost-effective to run a more thorough search, and file a formal patent application (utility or PCT).

B) Tailor The Searching Effort To The Purpose

The first rule is to tailor the searching effort to the needs of the company. Below is a diagram of a broad patentability search, in which the arrows represent searching efforts going off in many different directions, and the squares represent the patents or other documents you are seeking. The search here is wasteful because it goes off in way too many directions (lack of focus), and several of the searches go deeper into the field (are much longer) than is necessary.

Wasteful Search

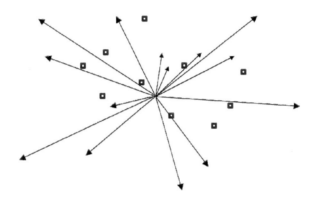

Figure 12

If the goal is to find out whether an idea is worth pursuing, the better approach is to more narrowly tailor the search. The reason is that the searcher need not find all the prior art to decide that an invention is unpatentable. *Whether the searcher finds one instance of the invention in the prior art or ten instances, the search is over.* Spending less on a focused search frees up additional funds for other searches.

Sometimes A Simple Search Will Suffice

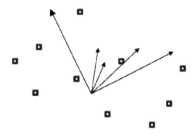

Figure 13

C) <u>Use Iterative Searching</u>

Probably the best strategy is to run your searches iteratively, i.e. in stages, with the most common type of iterative searching being iterative keyword searching. The first step is to use very specific keywords in a proximity-based system to see if there is an easy match. As an example, when searching for a screw with interleaved threads, I would search for a patent where the terms "screw", "inter*", and "thread" are all in the same sentence, or within 10 words of each other. If you find just what you are looking for, then the search is over. If you don't, then you need to keep broadening the scope (usually by removing keywords from your search) until you find at least a few patents or other prior references that match. This is where Internet searches often fail. Some Internet search engines (such as Google™) can handle wildcards, but they cannot handle proximity searching. One usually needs to use a paid service such as Lexis or Westlaw to perform proximity searching.

The second step in iterative searching is to examine the documents you found in the first step, and use them to refine your keywords. People use a great many terms and phrases to describe the same thing. A sawhorse may be described in one reference as a sawhorse, but in another reference as a cutting or sawing table. Similarly, the Internet may be alternatively described as a global network, or as a package switched network. You should look in each new reference for keywords that you hadn't thought of in the first place. In addition to identifying new words, you may even realize that you need to search for plural terms as well as singular terms. By refining the keywords your searches are both broader (in terms of using more alternatives) and more focused (because you can combine more search terms into a given search.

Iterative Keyword Searching
Focuses The Effort

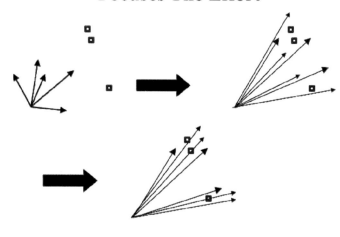

Figure 14

An alternative to iterative *keyword* searching is to use iterative *parent/child* searching. Here, the second step is to examine the documents you found in your first search, and use them as the basis for your next search. If you found patents in step one, then look at all the patents that are cited by the ones you found (children) and all the patents that are cited by the ones you found (parents). That second generation search may widen the catch to 40 or 50 references. Examine all those references to find the closest ones, and then look at children and parents of that second generation of references to find a third generation. Repeat that process until you either find just what you are looking for, or until the subsequent searches find only references that you already know about.

Iterative Parent/Child Searching Is Also Extremely Effective

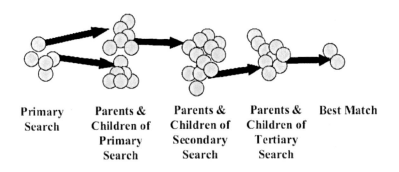

Primary Search	Parents & Children of Primary Search	Parents & Children of Secondary Search	Parents & Children of Tertiary Search	Best Match

Figure 15

D) Don't Allow Searching Costs To Get Out Of Hand

Many patent firms and independent searching services charge a standard, fixed fee (usually about $750) for patentability searching. That is a mistake. Fixed fee searches can make sense, but a one-size-fits-all standard fee can't possibly work well across completely different technologies. Some inventions are simple to search and others are notoriously expensive to search. A $750 fee, for example, overcharges for the simple searches and provides far

39

too little funding to adequately run complicated searches.

Patentability searches can cost anywhere from about $500 to $10,000 or more to accomplish what you want. At the low end patentability searching can be very inexpensive because it takes only a short time to find that the invention is already known. The better (i.e. more cost-effective) practice is to run those searches by the hour. If the very same idea can be located in 10 minutes, there is no need to complete the search and write up a formal report. The idea is dead, and the inventor should move on to something else. In our office we often run quick initial searches for free, and only charge if we need to move on to a more substantial search.

At the high end, patentability searching can be very expensive because of the databases used. For example, chemical inventions often require searching of chemical structures (as opposed to merely chemical names). There are only a few such databases available, and they can charge a small fortune. *Patentability searches can also be extremely expensive for simple inventions directed to common items.* Novel baby bottle designs, improvements to doorknob latches, new pens and pencils, and other very common items can be notoriously difficult to search. Such inventions usually have an enormous amount of prior art, and it is extremely difficult to sift through all that prior art. Simple inventions are also usually described with common language, so that a word search on a typical database can yield tens of thousands of hits. Still further, simple inventions can be difficult to search because the relevant products are often not patented. One needs to search way beyond the patent literature to general commercial and retail advertising.

Don't put too much emphasis on searches by technology area. I have approached searches using class and subclass limitations before, but I don't find them to be especially useful. All too often a technology of interest spans so many classes and subclasses that there is little or no benefit to identifying the relevant areas. Currently, just about the only time I use classes and subclasses is in limiting on-going searches, where there would be too many references without the limitation.

One important caveat is that one needs to use multiple sources, not just issued patents. Published patent applications, whether in the US, EPO, Japan, or elsewhere, can all be used as prior art to preclude patentability. Journals and other non-patent publications can also be used against a patent application. For simple, consumer products, I often ask my wife to search her catalog collection. Often an "invention" is already in the Hammacher Schlemmer ™ catalog! In fast moving areas such as electronics, software, and even mechanical arts, the Internet can also provide extremely valuable information. Recently we located a supposedly novel type of screw fastener being sold on the Internet. Apparently the product had never been patented. A good search involves at least several completely separate avenues of inquiry.

E) Use The Most Cost-Effective Sources

No matter what level of search you decide upon, and what strategies you decide upon, it is extremely important to choose the *right* information sources. One of the best free sources is probably the European patent office (http://www.european-patent-office.org/). The website includes access to US, European patents and applications, as well as PCT (WIPO) applications. The site also includes numerous links to Japanese and other government patent sources. The website of the US patent office (www.uspto.gov) is reasonably convenient for quick searches, and is very convenient for copying the full text of a patent application into a word processing document.

Use The Right Tools For The Search

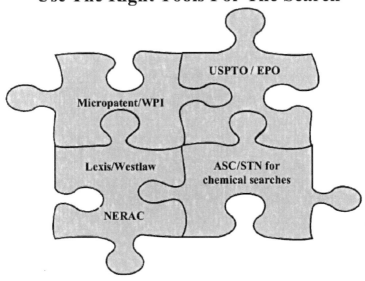

Figure 16

For patents other than US, Europe, and Japan, one of the best free services is Mayall's IP links at www.mayallj.freeserve.co.uk/front.htm. Among other things the site provides links to searching patents from many foreign countries, including Australia, Brazil, China, Hungary, India, and New Zealand. See www.mayallj.freeserve.co.uk/search2.htm#pa. The same site also has links to specialty databases such as the European coatings database, a fullerenes patent database, a DNA patent database, and even a saxophone patents database.

One of the best pay services is Micropatent™ (www.micropatent.com). For about $100 per day anyone can conduct full text searching on numerous databases, and secure patent family listings through INPADOC to boot. The Micropatent search engine is fairly powerful, although the site can be awkward and frustrating to use. Micropatent is fabulous for trademark searching as well.

For years we used a service called World Patent Index™ (WPI), available through Dialog (www .dialog.com) and several other on-line sources. The chief benefit there is that instead of searching patent text drafted by the patent attorneys, the system searches abstracts that are prepared by people knowledgeable in the field. That method goes a long way towards minimizing differences in terminology employed by different patent attorneys. It also tends to minimize the obfuscating effects of sometimes overly obscure patent drafting. Other significant benefits are that WPI provides patent family listings, on-going search capability, and even statistics on patent issuances by subject matter, inventor, assignee, and so forth. The biggest drawback is high expense. The service charges for computer time spent on each search, and then a per-record charge for data accessed. There are several strategies for minimizing those charges, and the customer service people are extremely helpful in assisting users to achieve cost-effective searching.

For chemical structure searches one almost has to use STN (http://www.cas.org/stn .html), or some other service that searches chemical abstracts of the American Chemical Society (ACS), http://www.cas.org/. Users generally cannot access the database directly, and instead work with a specialist to prepare the search. Companies that need to do a lot of chemical structure searches should secure a direct account with STN.

Probably the best source for accessing journals, newspapers and other publications is NERAC™. For a fixed annual cost of a few thousand dollars, NERAC provides unlimited access to tens of thousands of publications, as well as worldwide patents and published applications. Searches against the databases cannot be performed directly by NERAC customers, but instead are performed by extremely experienced professionals in all the different fields. Thus, a pharmaceutical search may well be performed by someone with an advanced degree in microbiology or biochemistry. Moreover, the person doing the search is available by telephone or email to discuss search strategy, and to fine-tune subsequent searches. Turnaround is usually a day or so, although NERAC is very good at handling emergency searches. NERAC is also wonderful for on-going searches through their on-line TechTrak facility.

LEXIS/NEXIS™ and WESTLAW™ are useful for searching US patents, especially for "right to use" as opposed to patentability searches. Both systems have very sophisticated proximity term search engines that can be used to select patents in which the search terms appear in the claims. Users can then download into a single file, segments of all the patents of interest (patent number, title, date, priority date, claim language), without downloading the entire patents. The downloaded file can then be searched and manipulated off-line using WORD™ or other word processing program.

I almost never rely on the results of outside searching services. There are numerous such services, especially in the Washington, D.C. area, and over time I have used several services that were recommended to me by one attorney or another. But without exception I have found the results to be unacceptable. For one thing, about 20% of the searches I do result in a critical "hit" within about 10 minutes of time. In those instances it seems silly to charge the client $750 or more for an outside service. My hourly rate is much cheaper. In other instances the search justifies the cost, but the searcher gains most of the insight resulting from the search, not the person drafting the patent application. Indeed, much of the benefit of searching lies not so much in identifying prior art, but in gaining ideas about what is known in the field, and how the claims of a new application could be written to take advantage of loopholes in that knowledge. Remember, we are trying to write claims based on a market-centered approach, not an invention-centered approach.

There is one outside searching service that provides stellar results. The company is Global Prior Art, Inc. Contact information is available on their web site at www.globalpriorart.com. Our preferred contact is Howard Davis. Global Prior Art's fees range from $5,000 to over $35,000 per search, but they pull out all the stops. They typically have PhD level people performing the searches, and have native foreign language speakers combing through the foreign prior art. They go way beyond ordinary key word searching, and way beyond the patent literature.

F) <u>Using PCT and EU To Perform The Search</u>

The reality is that cost is a key limiting factor in performing patentability searches. Some technology areas in particular are notoriously expensive to search. Chemical abstracts searching, for example, can cost upwards of $10,000 for some projects. Even moderately complex chemical searches can cost more than an inventor is willing to pay.

What one really needs is the experience of someone who has dealt with the particular technology in question for many years, and has the critical information at his/her fingertips. Even more, one needs the searcher to have unlimited resources, and not be limited in any way by costs. Well that describes patent examiners. Wouldn't it be wonderful if we could have the patent examiners do our search for us?

Well, we can do just that. The secret is to do a preliminary search, draft a decent patent application, and then file that application through the US receiving office of the PCT, while filing exactly the same application as a provisional with the USPTO. The examiner is supposed to issue a search report in about 16 months, but often does so in only 3 - 4 months. If the specification appears to have been reasonably crafted in view of the art cited by the examiner, the patent attorney can just proceed to prosecute that application. If, however, the PCT examiner cites references that are really problematic, then the attorney can abandon the PCT application, rewrite the previously drafted application to take into account those references, and then file a new PCT application, claiming priority to the provisional. No one will ever know that there was a previous PCT application. In extremely important cases, I have even filed a second PCT application right off the bat - but in the WIPO receiving office rather than the US receiving office. That parallel application is examined by the European Patent Office, which almost invariably identifies different prior art from that identified by the US receiving office.

A little known service of the European Patent Office is that they will conduct a search on a given topic for about $2,000, based upon a short disclosure of perhaps a paragraph or two. A top EPO official has assured me that such searches are given the same degree of care as patent applications, which means that the applicant is receiving probably the best searches in the world. The results are usually returned within about 6 months. The EPO will do both standard searches (which are the same as the prior art searches performed on official European and international applications) and special searches (which are "right to use" or "free exploitation searches"). A PDF downloadable document explaining these services is available at http://www.european patent office.org/dg1/brochures/index_search_doc.htm.

G) Ongoing Searches

Even very good searches can only identify materials that are published as of the date of the search. With technology moving forward at an incredible rate, every search becomes dated as soon as it is completed. The answer is to have on-going searches for topics of interest. As mentioned above, this can be accomplished through NERAC at zero marginal cost, or through other services for additional costs. In addition to using keywords, I recommend that on-going search strings include the patent numbers of relevant patents, and authors/inventors of interest.

Chapter III - Make Good Decisions On When And What Kind Of Application To File

Most people file a utility application first, followed by a PCT (international) application within the following year. Unfortunately, that strategy often results in long, drawn-out prosecutions, in which large sums of money are spent even before the inventor learns whether the invention is patentable. There are far better strategies. One strategy that is often very effective is to file the PCT application first, effectively naming the United States as a foreign country. If all goes well the inventor can receive the first office action in about four months, a US national phase application can be filed from the PCT, and the US patent can issue in about 14 - 16 months. Another good strategy is to file the US utility application first, but accompany the filing with a Petition To Make Special that advances the application to the top of top of the examiner's workload. Provisional patent applications can also be used creatively to dramatically reduce costs, and provide stronger patent protection.

A) <u>File PCT First</u>

When people think of a patent application they usually think of a utility application. There are other types, design applications for the ornamental appearance of objects, and plant applications for non-sexually propagated plants (e.g. roses), but those are not of concern here.

A utility patent application is normally the first filing that an inventor makes. Starting at month zero (as shown in the chart below), it takes about 1.5 years before the patent office comes back to the applicant with a first substantive office action. Yes, sometimes the patent office works faster, especially in technology groups that are not terribly busy. We once had a first office action on a toaster patent about two months after we filed it. But that is extremely unusual. At the other extreme office actions in the Internet and computer software fields can take 4.5 years or longer before the patent office issues its first office action!

Standard Approach To Utility Applications

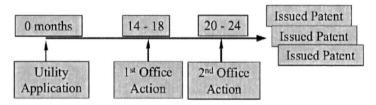

Figure 17

The first office action is almost always a rejection, as it should be. If you offer to sell your car for $30,000 and someone immediately pulls out his checkbook to write the check, the first thing that should go through your mind is that you offered to sell the car at too low a price. The same is true of the patent office. If the office accepts all the claims on the first go-round, the patent attorney and the applicant should seriously consider whether the claims were too narrow.

Assuming that the first office action is a rejection of at least some of the claims, the patent attorney then argues back and forth with the patent examiner. That process usually takes about six months to a year, and usually involves a second or even a third office action. Assuming all goes well, the patent can actually issue in about two and a half to three years. If there are divisional applications, they can issue in about the same time if the paths are parallel, or they can issue much later (many years later) if the paths are consecutive.

Foreign filing must be done within one year of the earliest claimed priority date of the utility application. If the utility application is filed first, then the foreign filing must be done by the one year anniversary of the filing of the utility application. Foreign filing costs about $4,000 to $5,000 per country or region to file, so the costs can add up pretty quickly.

One way to forestall paying all those foreign filing charges is to file a PCT (Patent Cooperation Treaty) application, designating the foreign countries of interest. In that respect PCT can be thought of as a doorstop that keeps the door open for up to 30 months for foreign applications to claim priority back to the US application. The PCT application must be filed by the one year anniversary of the earliest claimed US application, which in our example is the utility application.

The PCT filing is beneficial for other reasons as well. For one thing the PCT generally does a better search than the US patent office. Even though the receiving office for the PCT in the United States is the USPTO, searches performed for a PCT application are "international type" searches, and are supposed to be better than the ordinary searches.

Another benefit is that the arguing period is generally shortened. Under the new rules adopted by the PCT as of January 1, 2004, the PCT is supposed to issue a combination search report and written opinion by the 16th month after the earliest claimed priority date. There is no longer any formal provision by which the PCT is required to argue back and forth with the applicant or applicant's attorney, but when such arguing does take place, it should be relatively compressed just as it was under the old rules. The reason is that applicants need to push forward with their foreign filings, and the PCT tries to accommodate that need. If arguing does take place the examiner will issue a preliminary examination report, and the attorney/applicant moves on to national phase filings. The process is depicted below.

Standard Approach To Foreign Filing

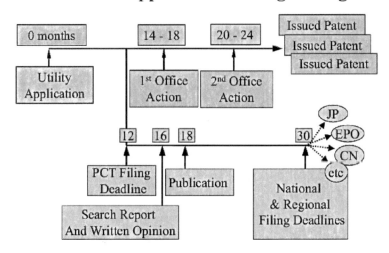

Figure 18

The problem, of course, is that since a PCT application must be filed within one year of the earliest filed US application, the US patent office will not have issued the first office action (search report) by the deadline to file the PCT application. Thus, when the PCT filing deadline rolls around, the attorney/applicant doesn't know whether it is worthwhile to file foreign or not.

There is a better way. It is perfectly legal to file the PCT application first, which in effect names the United States as a foreign country. In that case the PCT still has 16 months to issue the search report, but if history is a guide many of the examiners will issue that search report / written opinion about three and a half to four months! That is a huge advantage when trying to raise money from the marketplace in the early life of a company. Investors are understandably skeptical about what the inventors says about patentability, and even what the patent attorney says about patentability. But here we have a way to secure a search report from the patent office itself, in a matter of months instead of years.

File PCT First (Better Approach)

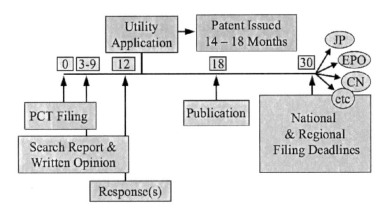

Figure 19

There are other advantages as well. If the examiner does respond quickly with the combination search report/written opinion, and can be persuaded to argue back and forth with the attorney or applicant, and if the applicant pays the Chapter II fee of about $700 when he files the response to the search report / written opinion), then the entire process of arguing back and forth with the examiner can be completed near the end of the first year. The attorney/applicant can then file the US national phase of the PCT application, and if all claims were deemed allowable at the PCT stage, the US application will likely not be examined any further, and will go right straight off to issuance.

Now it is true that some examining groups (e.g. the Internet and software groups) re-examine everything, even if all claims were deemed allowable at the PCT stage. But that is the exception rather than the rule. Indeed, when filing the US national phase on an application with all claims allowed, the filing fee is only about $100, because the patent office is not contemplating an additional search. If all goes well, the US patent can issue within the same year and a half period that would be required just to get the search report under the normal filing procedure.

Yet another advantage of filing the PCT first is that the PCT deals with divisionals differently from the US patent office. Instead of issuing a "restriction requirement" to split the application into multiple parts, the PCT issues disunity rejections that can usually be resolved by filing an extra $240 fee for each separate "invention". Upon filing national or regional applications at 30 months, the various countries (including the US) often keep all the claims together rather than issuing restriction requirements - especially if all claims were deemed allowable by the time the preliminary examination report was issued. *As discussed above, that can easily save hundreds of thousands of dollars.*

There is even a further advantage in filing PCT first, namely simplification of the US prosecution.

Amendment of claims during US prosecution gives rise to "file wrapper estoppel", which precludes the applicant from later arguing that the issued claims cover subject matter that was given up during prosecution. For awhile the law was even stronger, precluding application of the doctrine of equivalents (which broadens the scope of patent claims beyond their literal meaning) for any claim that was amended in any manner. The US Supreme Court overturned the complete bar of the so-called Festo doctrine, but there is still a very strong presumption that any amendment made to the claims during prosecution eliminates the ability of the patent holder to later rely on the doctrine of equivalence with respect to the element that was modified. In any event, the ability to resolve many or all of the objections and rejections during the PCT phase of prosecution means that there is much less amending of claims during the US national phase - and that can do nothing but help.

Filing PCT first also opens up a very interesting strategy for using provisional applications - something that could be called "patent judo" since you use the examiner's own arguments against him. As shown below, you can prepare and file a PCT application, and then photocopy the application and file it again on the same day as a provisional application. If the patent examiner comes back with a terrible search report, you may wish that you could start all over again with a new application. Using this approach you can do that. All you need to do is abandon the PCT application, and draft a new PCT application taking into account all the prior art that the examiner found. You then file that new PCT application claiming priority to the provisional filing, so that no time was lost. Since the examiner already took his best shot at you, the new PCT application can be written to overcome the examiner's arguments. Voila! The claims should be deemed allowable, and the utility application should issue in good time.

File PCT First (With Provisional Filing)

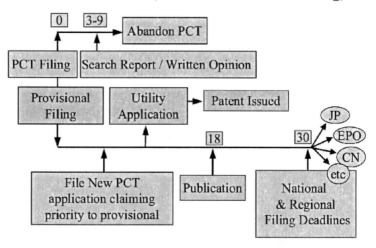

Figure 20

Of course you can do the same thing with a utility application in place of the provisional if you want. The strategy works as long as there is an existing priority application to which you can claim priority when filing the second PCT application.

B) Consider Filing A Petition To Make Special

Another great strategy is file a petition to make special when filing the utility application. There are currently 12 special circumstances under which the patent office will push a patent application to the top of the heap. (See 37 CFR 1.102 Advancement of examination, and MPEP § 708.02 Petition To Make Special). Those are (1) prospective manufacture; (2) current infringement; (3) applicant in poor state of health; (4) applicant at least 65 years old; (5) inventions that materially enhance the quality of the environment of mankind by contributing to the restoration or maintenance of the basic life-sustaining natural elements; (6) inventions that materially contribute to (A) the discovery or development of energy resources, or (B) the more efficient utilization and conservation of energy resources; (7) inventions relating to recombinant DNA; (8) the applicant submits a statement that a pre-examination search was made, and complying with certain formalities; (9) superconductivity materials; (10) HIV/ AIDS and Cancer; and (11) countering terrorism; (12) biotechnology filed by small entities.

Although many of the possibilities are very specific, the eighth one can apply in almost every instance. Indeed, a patent attorney/applicant cannot write an effective patent application without knowing the extent of the prior art - and that means running an adequate prior art search. Indeed, the eighth possibility can be used to great advantage when the applicant uses the PCT first filing strategy. In that case the applicant can file a petition to make special when filing US national phase, based on the PCT search, even if the preliminary examination was not favorable. The applicant need only distinguish the remaining references, which is something he has to do anyway when filing the preliminary amendment in the national phase.

A petition to make special speeds up patent prosecution considerably. Instead issuing a first office action at the usual 18 months, the patent office will often grant the petition to make special within about two months, and then issue the first office action within about four or five months. Indeed, even subsequent arguing back and forth between the examiner and the patent attorney/applicant is speeded up, so that second and third office actions are issued more quickly as well.

Petition To Make Special Speeds Up Patent Prosecution

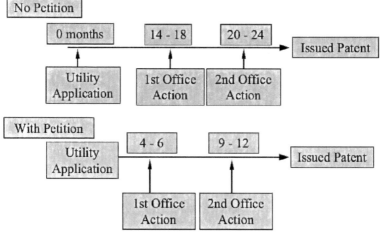

Figure 21

The key is to run a decent patentability search, draft the claims to circumvent all the prior art found during the search, and then prepare a short analysis describing how the claims circumvent the prior art. But this should not be particularly burdensome. It should have been done anyway as a matter of good patenting practice!

One very interesting side note is that petitions to make special can be effectively combined with PCT applications. If the patent attorney/applicant filed PCT first, he can file the US national phase with a petition to make special as soon as he receives a favorable search report, written opinion, or preliminary examination report. This forces the patent office to consider the application sooner, and the patent office can hardly reject the petition to make special on the grounds of inadequate search. Their own people did the search!

C) <u>Use Provisional Applications Effectively</u>

Provisional patent applications, which are commonly known as "provisionals", are a wonderful development. They don't require the attorney or applicant to draft any claims, they are never substantively examined, and they never issue as a patent. Provisionals are also inexpensive to draft and file, and they serve as a placeholder to which an inventor can claim priority in a subsequent formal application (utility or PCT).

In addition to describing an invention, i.e. describing something that is new and non-obvious, provisionals have only two basic requirements; (1) they need to enable the invention (explain to someone of ordinary skill in the art how to make and use the invention), and (2) they need to satisfy the "best mode" requirement (disclosing the best way known to the inventor as of the filing date for making and using the invention). Assuming those requirements are satisfied, a provisional can be prepared and filed in less than an hour. I have personally filed provisionals that contain nothing more than photocopies of lab notebooks, and I have filed provisionals that are little more than a drawing and a single explanatory paragraph. I once even filed a provisional with nothing more than a drawing! The filing was immediately rejected by the patent office as being inadequate, but I prevailed by demonstrating that the two basic requirements of enablement and best mode were adequately satisfied by the drawing.

The filing fee for a provisional is quite low, only about $75 for a small entity (generally less than 500 employees) and $150 for a large entity (generally 500 employees or more).

Surprisingly, many patent attorneys make little use of provisionals. One problem is that it is difficult to charge much money for a provisional. After all, how much can one charge for slapping a cover sheet on a disclosure, and filing it as a provisional? Granted, it is highly desirable to modify the disclosure to some extent, to broaden the language and stop the inventor from saying something that shoots himself in the foot. But even so, how much can one charge for an hour or two of work? From a patent attorney's standpoint, it is much better to draft and file the formal application.

On a less cynical note, there are potentially valid reasons for being leery about filing provisionals. One disadvantage is that they die after exactly one year, with no possibility whatsoever of extending them. But that disadvantage can be overcome by filing a formal (utility or PCT) application during the one year life of the provisionals. It is also true that a priority claim back to a provisional is only valid to the extent that the claims filed in the later application were adequately disclosed in the provisional. This can be a real trap for inventors, because they tend to believe that filing a provisional "covers" the invention. The provisional may indeed cover the invention, and then again it may not, depending on the depth of the disclosure in the provisional.

Another potentially valid reason for avoiding provisionals is that filing a provisional only delays (rather than eliminates) the day of reckoning, when the inventor/applicant must pay the patent attorney to draft and file the formal application. If the money must be spent anyway, perhaps it better to spend it now, file the formal application, and get the clock ticking on the first office action.

Assuming one is contemplating filing a provisional, how should that filing fit into the overall patenting strategy? There are several important ways to use provisional applications, and not all of them are completely obvious. The normal use of a provisional application is as a precursor to a utility application. In the diagram below, a single provisional is filed at 0 months, and expires at 12 months. As already discussed, the provisional can be very simple and inexpensive, or it can be drafted as a complete formal application and merely filed as a provisional to delay filing the formal application. Of course the drawback is that the patent office tries to get out the first office action at about 14 - 18 months after the US application is filed. Delaying filing of the utility by filing provisionals first pushes back the date when the clock starts ticking for the office action.

In any event, in order to take advantage of the priority date of the provisional application, the applicant must file a utility or PCT application claiming priority to the provisional before the provisional expires at the end of the 12 months. In the diagram shown below the patent attorney/applicant filed at about eight months into the twelve month period. If the inventor is unlikely to develop any new subject matter, and especially if the provisional is relatively thin, it is desirable to file well before the 12 months is up.

Normal Filing of A Provisional Application

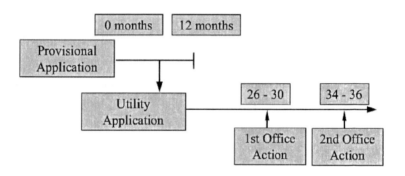

Figure 22

This "normal" usage works very well for circumstances in which there is only one invention, and the inventor/applicant has a clear understanding of what the invention is at the time the application was filed.

We have also used single provisionals to file multiple inventions. The most common occurrence is to write up a few pages of description that disclose the various inventions, and then pull out the various pieces as needed when it comes time to write and file the formal applications.

From time to time we have also written several full utility or PCT applications, but then for various reasons the inventor decided not to spend the money right away to file the formal application. In those instances we simply stapled all the various applications together and filed them all as a single provisional. Then, before the one year deadline, we separated out the formal applications again, and filed them again as individual applications.

Several Inventions Can Be Combined Into A Single Provisional Application

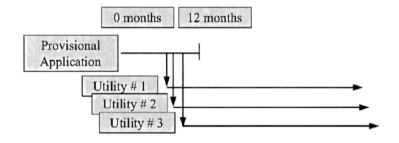

Figure 23

Provisionals are especially useful where the inventor expects to develop some new wrinkles on his invention after we file the first application. One can't file a second provisional claiming priority to the first provisional, but one can serially file multiple provisionals, and then have the formal application claim priority to some or all of the various provisionals filed within 12 months. In the following diagram, the utility claims priority to five different provisionals.

Single Utility Can Claim Priority To Multiple Provisional Applications

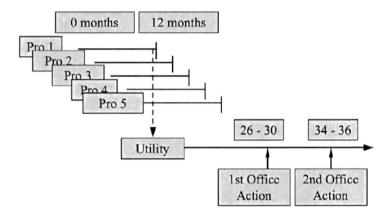

Figure 24

Another great use of a provisional application is to provide a quick and inexpensive priority date for a C-I-P (continuation in part) application, where the applicant files a new application adding additional disclosure. For example, if a utility application was filed with disclosure to several embodiments of an invention, and then after the utility was filed the inventor devised yet another embodiment, the inventor could file the new material in a provisional, and then file the C-I-P application claiming priority to both the utility and the provisional. The C-I-P application would then have two priority dates, one

70

for the material disclosed in the utility application, and one for the new material disclosed in the C-I-P application. Indeed, the C-I-P can claim priority to the utility and multiple provisionals as shown below.

Provisional(s) Can Be Used As Precursors To Filing CIP Application

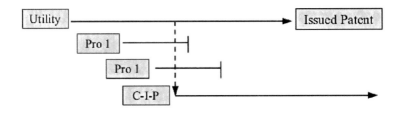

Figure 25

Chapter IV - Write The Specification As A Sales Pitch

A patent application consists of a title, specification, at least one claim, abstract, and a drawing. The specification is often 10-20 pages long, and drafted to an engineering audience, from a dry, scientific standpoint. Big mistake! The specification should be written in a concise manner, from a marketing/sales viewpoint, clearly identifying what is new and exciting about the invention. After all, the invention will only get patented if the examiner can be convinced that the invention is an improvement over the prior art. Like any good sales document, the specification should describe the problems being addressed, describe why the prior solutions are not completely satisfactory, and then describe how the present invention solves those problems.

The Background is an especially good opportunity to show the patent examiner (and later on a judge or jury) that the claimed invention really is a significant advancement over the prior art. From that perspective one of the main purposes of the Background section is to identify a problem, demonstrate how difficult it is to solve that problem, and how brilliant (or clever, diligent, observant, or whatever) the inventors were in solving that problem. That approach automatically places the claimed subject matter in the most favorable light to secure a patent, and yields all sorts of support that can be used later on during prosecution to argue against obviousness rejections. It also helps later on during litigation.

The sales aspect of the Background can be achieved most effectively by clearly identifying a problem, describing the various ways that others tried to solve the problem, and then discussing why those previous solutions were inadequate. That process is shown below.

The Background Should Show How The Problem Is Intractable

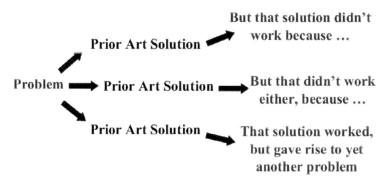

Figure 26

The "problem to be solved" is best framed in such a way that either no one else clearly identified the problem before, or if the problem had been previously identified, no one else was able to solve it.

That task is not so difficult as it may seem. The framing step is just a rearrangement of the wording of the first paragraph of the Summary. In US 6272727, for example, the Summary section begins:

> It has recently been discovered that the above-mentioned problems can be resolved by biasing the upper portion of files in a file drawer or other file holder towards the rear of the file holder.

Rearranging the above sentence, we can recite the problem as a "need for devices and methods which can conveniently position file folders . . . ." Some readers will recognize this approach as being analogous to that popularized by the Jeopardy™ television show. - use the answer to define the question.

Once we have a statement of the problem to be solved, we can then go back and build a case for why that problem has been so difficult to solve. A general overview focuses the reader's attention on the environment in which the problem occurs.

> Folders are ubiquitous in modern offices. In most cases folders comprise little more than a folded piece of paper or plastic, with the

"fold" ranging anywhere from the hard, creased fold of a standard "manila" folder, to a soft "fold" or bend of a hanging folder. Folders generally also have a tab or other extended portion at one of the sides for including identification information. In most instances office folders are used to contain papers, and where the papers comprise a file of information, the folder is properly called a file folder. But it is widely accepted that items sold as file folders may also contain computer disks, writing implements, and a host of other items, and the terms file folders and folders are used synonymously herein in a broad generic sense.

Next, we describe how the problem manifests itself.

Folders are generally intended to be kept upright in a file drawer, file box or other file holder so that the file identification information is viewable on all of the files at the same time, and so that each of the files are more or less accessible without significantly disturbing the other files. It is not always feasible, however, to maintain folders in an upright position, and this is largely due to the fact that smaller and heavier items tend to fall to the bottom (folded portion) of a folder, where they increase the thickness of the folder. Where this occurs in multiple files within the same file drawer or other holder, all of the folders tend to fall forward or backward in the file holder. While this problem is often only a minor nuisance, it does create difficulty in viewing the folder identification information, and in accessing individual folders and their contents.

We then follow up with descriptions of the various ways others have failed to solve the problem.

The problem has been addressed in part by providing a moveable partition. Such partitions can effectively reduce the volume of space available to the files, thus tending to keep them upright. Alternatively, a partition can be made to pivot against the floor of the file holder. This biases the bottom portions of the files forward and allows the top portions of the files to fall backward. While such partitions are of some use, the relatively free pivoting tends to tilt the files much farther than is necessary. Previously known such partitions are also limited to file drawers, such as those found in metal cabinets, which are strong enough and otherwise adapted to securing such partitions.

Then Describe The Ideal Solution As What The Invention Provides

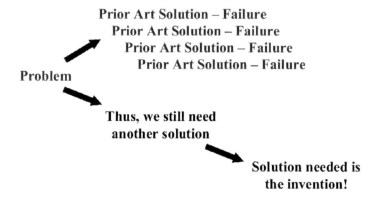

Figure 27

The final step is to restate the problem such that the claimed invention exactly fits the bill. This is, of course, exactly what a good sales presentation is supposed to do; describe a problem to which the salesman's product is a perfect solution. To this end, the Background section in the example above ends with the following statement: "Thus, there remains a considerable need for devices and methods that can conveniently position file folders in substantially any file holder."

Chapter V - Write The Claims Using A Market Centered Perspective

The vast majority of patent attorneys writes claims from an " invention oriented" perspective. That approach can work well when the invention is a dramatic improvement over the prior art, but is a terrible strategy for the vast majority of inventions that are merely improvements over what is already known. Instead of writing technology driven, invention-oriented claims, patent attorneys should write market-oriented claims that focus on keeping the competition out of a given marketing space. The market-oriented approach provides much, much broader protection for the patentee.

A) Claim Everything That Is Available

Every invention is really a solution to a problem. Ideally, the inventor would claim every commercially viable solution, which in the following diagram would be the entire space.

An Invention Is A Way To Solve A Problem

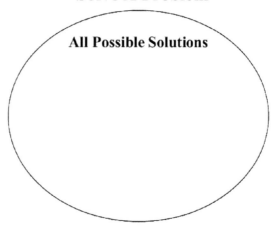

All Possible Solutions

Figure 28

But that just isn't possible. Almost every inventive space is already occupied by at least some prior art, which acts as a sort of land mine preventing the inventor from capturing the entire space.

But Some Solutions Are Already Known

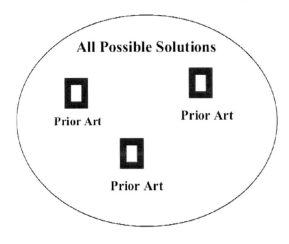

Figure 29

In the diagram shown below, the invention is shown as a light bulb. One could, of course, claim the area occupied by the light bulb. But that is usually a very bad idea because the "invention" as contemplated by the inventor, covers such a small area.

Invention Centered Approach
Focuses on Claiming The Invention

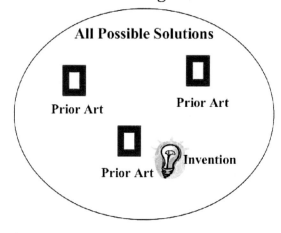

Figure 30

Indeed, new patent attorneys are taught to look beyond the inventor's *preferred embodiment*, to claim the *concept* of the invention as broadly as possible. This is what patent attorneys mean when they say they file the broadest possible application. The obvious limitation, of course, is that the claims can never be so broad that they include what is already in the prior art. The largest area that can be drawn from the [invention] light bulb is shown below, because it sits right up against one of the prior art references.

Broadest Conception Of Invention
Runs Up Against The Prior Art

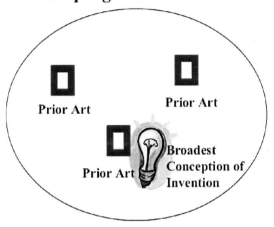

Figure 31

But even that supposedly broadest possible claim "leaves a lot money on the table. "The much better approach is to focus on the marketplace rather than the invention. Lets start with the same conceptual space, and the same invention.

Market Centered Approach
Starts With Same Invention

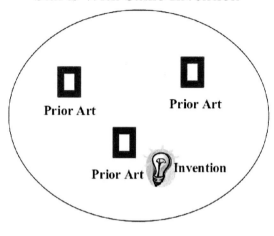

Figure 32

In this case the patent attorney works with the inventor to figure out what is left open in the field, i.e., we strive to identify what other inventions are needed to monopolize the remainder of the market. By appreciating what is remaining, the patent attorney and inventor can conceptualize other inventions.

But Focuses On Alternative Solutions

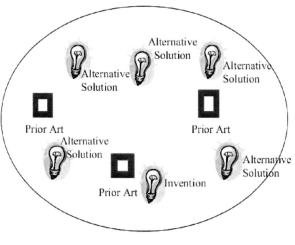

Figure 33

Does that make the patent attorney a co-inventor? No, of course not. The fact is that if the patent attorney can devise all these other solutions from the one solution developed by the inventor, then the other solutions were inherent in the inventor's invention.

For example, years ago everyone knew that having pop-tops pop off soda and beer cans was a bad idea. The tabs cut people's fingers, caused pollution, and even ended up being swallowed by fish. Then someone came along with the idea of a pop-top that didn't pop off. Yes, he had a preferred embodiment, but that wasn't the invention. The invention was not how to do it, but the idea of doing it! All those other solutions were inherent in the inventor's concept.

Indeed, all those other ideas needed to be claimed in order to cut out the competition. From a marketing standpoint, it really didn't matter how the inventor thought the new tab should be designed, as long as it kept competitors out of the marketplace. What needed to be claimed was all of the remaining space! That is what the patent attorney (and the inventor) should be focused on, not the invention.

The final step of a really good patent attorney is to do the difficult mental step of figuring out how to claim all those different inventions. It can be done! When the process is finished, a patent application is written that basically claims everything that is left in the space, i.e. all the space that is not covered by the prior art. The resulting coverage looks something like a golf course, with the claims covering all of the available subject matter.

Market Centered Approach Then Claims All Available Subject Matter

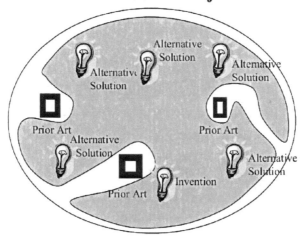

Figure 34

Chapter VI - Make Appropriate Use of Independent Claims

All patent claims are either independent (stand alone) or dependent (refer to another claim). Using a large number of independent claims is extremely wasteful, unnecessarily increasing the cost of the application and prosecution, up to three quarters of a million dollars! In addition, clever use of dependent claims can force a broad interpretation on the independent claims, and strengthen the patent considerably. Excessive use of independent claims is a hallmark of a poorly written patent application, and an attorney who failed to do the heavy mental lifting to properly understand the invention.

A) What Are Independent And Dependent Claims?

All patent claims must be either independent or dependent. Independent claims stand alone, and do not reference any other claim. Thus, if claim 1 recites "A chair having only two legs" as in the example above using the Venn diagrams, then claim 1 would be independent because it does not reference any other claim. Claim 1 is always independent.

Dependent claims reference another claim. Continuing with the example above, claim 2 would be dependent if it recites "The chair of claim 1, wherein the chair is made of metal." Dependent claims include all the limitations of the referenced claim. To infringe claim 2, a competitor's chair would not only have to be made of metal (the limitation stated in claim 2), it would also need to have only two legs (the limitation imported by reference to claim 1).

Claims can be dependent on an independent claim, or on another dependent claim. In this simple example, claim 3 might recite "The chair of claim 2, wherein at least a portion of the chair is covered with a fabric." Claim 3 would thus be limited to two-legged chairs made of metal, and covered with a fabric.

Just to complicate matters, claims can also be multiply dependent. Those claims are of the form, "The widget of any of claims 1, 6, or 9, comprising ...." Multiple dependent claims are extremely expensive, and very rarely worth the money.

B) <u>Why Not Have All Independent Claims?</u>

(1) <u>Independent Claims Increase Filing & Prosecution Costs</u>

Dependent claims are useful for many reasons. The most straightforward is that they are less expensive than independent claims. The standard filing fee for a utility application with the US patent office covers up to three independent claims and up to 20 total claims before additional filing fees are incurred. Additional fees for small entities (companies with less than 500 employees) are currently $43 per independent claim over 3, and $9 for each claim over 20. The fees are double that for large entities .

Those additional fees can add up quickly. Lets say a large company files a patent application with 20 independent claims. The government filing fees (for a small entity) are the standard fee of $385, plus the penalty fees for the 17 extra independent claims, which comes to 17 * 43 = $731, for a total of $1,116. Using all those independent claims almost tripled the filing cost.

It quickly gets worse. The first substantive action of the patent office will be to issue a restriction requirement, forcing the applicant to either abandon, withdraw, or divide out the various independent claims into perhaps five, ten, or even up to twenty separate applications. Lets say the applicant decides to pursue five separate applications. He now has to abandon or at least withdraw several of his originally filed claims with all the attendant prejudice resulting from that action, and in addition he must file another five divisional patent applications. The out of pocket costs on those five divisional applications is 5 * 385 = $1925, plus paralegal time to file the paperwork. The total cost of filing fees for the twenty original independent claims is now over $3,000 instead of the original $385. If this were a large entity (over 500 employees), the unnecessary additional US government filing costs would be over $6,000!

Even if the examiner doesn't issue a restriction requirement, it is unlikely that all twenty claims will be granted together in the same patent. The more likely result is that after prosecuting all those independent claims for a few years, the patent office will deem three or four of the claims to be allowable, and the applicant will pay the issue fee and have those allowed claims go to issuance. But that leaves seventeen non-issued claims, which must then be filed in a divisional application. The cost of that filing is $385 plus the penalty fees for the 14 extra independent claims, which comes to 14 * 43 = $602, for a total of $987.

Independent Claims Greatly Increase The Cost Of Securing US Patents

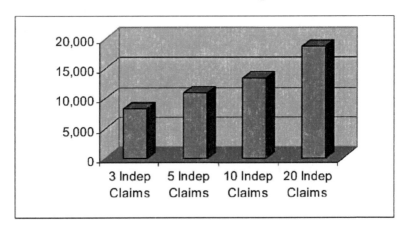

Figure 35

All of that is unnecessary. The application could just as well have been drafted with one or two independent claims, and the remainder filed as dependent claims. The problem, of course, is that the patent attorney who is trying to build up a big case load would only have one case to prosecute rather than five or six cases.

(2) Independent Claims Increase Cost of Foreign Prosecution

As described in Chapter I, foreign filing multiplies the problems of US prosecution. Where a single US application with a few independent claims may spawn a family of half a dozen or more foreign applications, the same US application drafted with many independent claims will likely spawn separate families of foreign applications for each of the US divisionals. This is a big reason that a single US application can end up costing the company a million dollars in prosecution fees!

Independent Claims Increase The Cost Of Foreign Patents Even More

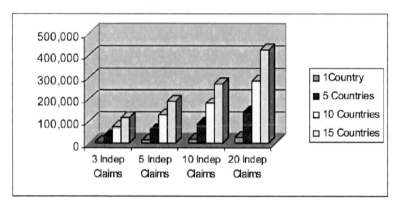

Figure 36

In addition, the patent offices of many foreign countries refuse to deal with multiple independent claims. Thus, even if the US patent office allows the applicant to keep several independent claims within the same utility application, a foreign patent office may still require the applicant to divide out the foreign application into several divisional applications.

Since the expense is often too high to file all those divisionals, the applicant is forced to re-draft his claims to be dependent upon a single independent claim. Not only is this additional work, but then all those new claims have to be translated. Well that just begs the question of why the patent attorney didn't draft the claims that way in the first place.

(3) <u>Too Many Independent Claims Shows Failure To Understand The Invention</u>

The added out of pocket costs of using too many independent claims is only part of the problem. The bigger problem is that filing a great many independent claims is often just a smoke screen for an attorney failing to do the heavy mental lifting of figuring out what the invention really is. The "value added" of a good patent attorney lies not so much in his knowledge of the field of the invention, or the drafting of application *per se*, but in his conceptualizing of the "invention" at an appropriate level of abstraction, such that the commercial value is clearly identified and claimed.

The key here is *reductionistic* thinking. The patent process can be thought of a double-sided funnel, such as that shown below. At the left side of the diagram the inventor presents the patent attorney with his invention. The inventor usually thinks he is quite clear about what the invention is, but his clarity is deceptive because he has almost always confused "the invention" with his "preferred embodiment" for carrying out the invention. In the middle portion of the diagram the patent attorney brainstorms with the inventor, and distills all of the inventor's ideas and embodiments down to a critical few elements of the "invention". That process of distilling the invention down to its critical essence is reductionistic thinking; something that very few people are good at. Once that is done, the rest is easy. The right side of the diagram shows that the attorney is expanding out each one of the critical elements into all of their commercially viable alternatives.

Symbolic Drawing Of A "Good" Patent

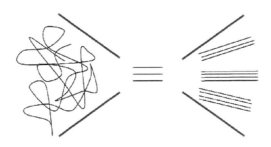

Figure 37

This process can easily go towards a "good patent" as symbolize above, or a "bad" patent as symbolize below. The difference is really the attorney's ability for reductionistic thinking. In the example above the attorney is almost perspicacious in his thinking, clearly drawing out the few elements that render the invention important.

In the example below, the attorney merely summarizes the idea without ever really determining the critical essence of the invention.

Symbolic Drawing Of A "Bad" Patent

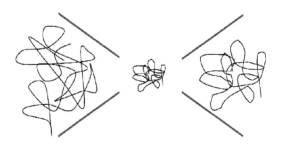

Figure 38

Despite all the advantages of filing only one or two independent claims, it is true that many patent attorneys like to use five, ten or even more independent claims. Indeed, some applicants equate a plethora of independent claims with a good application. Since a large number of independent claims usually leads to a large number of patents, those tend to be the same people who want a large portfolio of patents to impress other people. Well, there is a place for such thinking.

Unsophisticated investors often ask how many patents the company has, and potential employers often ask an engineer or scientist how many patents he has to his credit. They rarely look to the detail to determine how valuable the patents actually are. So it is not necessarily *wrong* to file applications with a large number of independent claims; it is just unnecessarily expensive. As long as these competing considerations are consciously weighed, a decision to go either way may be reasonable.

Chapter VII - Claim With Litigation In Mind

The main goal of patenting is to keep competitors out of what you consider to be your marketing space. This can sometimes be accomplished with threats and cajoling, but from time to time the only way to effectively keep competitors at bay is through litigation of patent claims. The litigation route is effective, however, only if the claims are broad enough to cover substantially all of the commercially viable alternatives. Otherwise the competitors simply design around the claims. The keys to writing broad claims are discussed below, including the use of short claims, use of the specification to define key terms tautologically, and use of overlapping target claims. It is also important to focus on claims that can realistically be enforced. Methods of treating illnesses, for example, are often unenforceable regardless of claim breadth. The patent holder cannot realistically sue the physicians or patients that are infringing the patents. Means-plus-function claims are also often unenforceable. They tend to be much narrower that the applicants think they are.

A) <u>Insist On Short Claims</u>

Many people think of patent claims the way they think of ordinary real estate property claims. They think that a long claim is like a bigger piece of property; it has as more "stuff" in it and is more valuable. But exactly the reverse is true. Every word in a patent claim is another limitation, and *the longer the claim is the narrower it is*. A shorter claim is almost always better than a longer one, and is more cost effective because it includes more subject matter.

As a first example, consider the first person who invents a two-legged chair. Yes, a two legged chair does exist.

Figure **39**

Lets assume that the inventor believes metal is the only material strong enough to support a two-legged chair, and he also thinks that metal chairs are only viable if they are covered with fabric. Thus, his broadest claim is "A two-legged chair made of metal and covered with fabric." That might seem like a nice, short, claim, but it is way too narrow as shown by the following series of diagrams. The set of all two-legged chairs is shown graphically below.

Two Legged Chairs

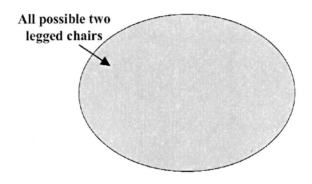

Figure 40

The set of two-legged chairs *made of metal* is smaller than that (a "proper subset"), and includes only the smaller circle in the diagram below.

**Two Legged Chairs
Made Of Metal**

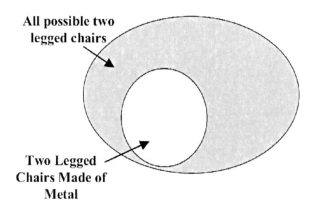

Figure 41

The set of two-legged chairs *covered with fabric* is also smaller than the set of all two-legged chairs, but in a different manner from two legged chairs made of metal.

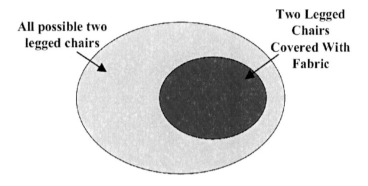

Figure 42

Finally, the set of two-legged chairs *made of metal and covered in fabric* is smaller still, because it only includes the overlap of the two subsets. Thus, a claim to a "two-legged chair made of metal and covered with fabric" is too long, and therefore too narrow. It contains unnecessary limitations. If the inventor was truly the first person to devise a two legged chair, the better claim would be to simply state "A chair having two legs."

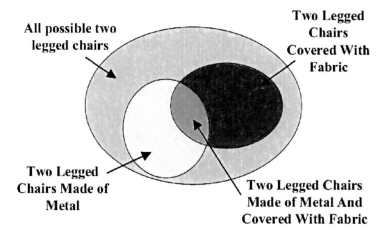

Two Legged Chairs Made Of Metal And Covered With Fabric

All possible two legged chairs

Two Legged Chairs Covered With Fabric

Two Legged Chairs Made of Metal

Two Legged Chairs Made of Metal And Covered With Fabric

Figure 43

B) <u>Omit Unnecessary Elements</u>

Omitting unnecessary elements is critical to drafting short claims. Consider, the very first person to invent a chair of any type. He claims "a chair having four legs, a vertical back, held together with nails, screws, bolts or dowels." That claim is way too narrow! The seat doesn't have to be flat, the back doesn't have to be vertical, the chair could be built with some other number of legs, and the seat, back, and legs could be held together with connectors other than those listed. The following chart shows how the original claim should be reworded to omit unnecessary elements.

All Unnecessary Elements Should Be Omitted

Example: world's first chair

A chair comprising:
- a flat seat;
- a vertical back;
- four legs; and
- the seat, back, and legs held together by nails, screws, bolts or dowels.

A chair comprising:
- a seat;
- a back;
- at least two legs; and
- the seat, back, and legs held together by connectors.

Figure 44

Everyone says that they eliminate unnecessary elements in patent claims, but in reality they don't do it. If you have filed patent applications in the past, go back through the independent claims and ask yourself whether every single word is necessary to claim around the prior art. If the answer is no, then the claim should probably have been simplified.

There are, of course, exceptions. Sometimes an invention involves so few elements that the simplest claim would be rejected by the patent office just by virtue of its simplicity. In that case it can be a very good idea to include a few other elements, just to make allowance more palatable to the examiner. But you should be aware of what you are doing; which is essentially putting up a smoke screen.

C) <u>Define Key Terms Tautologically</u>

In that last example we used the word "connectors" to replace "nails, screws, bolts, and dowels". But what does that buy us? Does the term "connectors" include all possible types of connectors, including glue? It does if the patent attorney defines his terms tautologically rather than by simply listing the known choices.

A tautology is a classification scheme that covers all possibilities. One of the main tricks of drafting patents is to list the known choices on a piece of paper, and then classify them in some manner. The list of "nails, screws, bolts, and dowels", for example, can readily be classified into those that are threaded and those that are not threaded. This is immediately a tautology because everything must be either threaded or non-threaded. However, abstracting just another step, we can see that both threaded and non-threaded connectors are examples of a broader class of mechanical connectors.

Abstracting still further, we see that a sister class to mechanical connectors are chemical connectors. Once we contemplate chemical connectors, we can then very quickly think of all the various types of chemical connectors, glues, adhesives, and so forth.

But we are not done yet. If there are mechanical and chemical connectors, there is probably a third category, which we can call non-mechanical and non-chemical connectors. The rule is that if we can think of a genus of that class, we can claim the whole class. In this case we realize that one could conceivably hold a chair together using magnets. Claiming this class provides us with the final piece of out tautology. From a purely logical perspective, all connectors in the universe must be (a) mechanical, (b) chemical, or (c) non-mechanical, non-chemical, because that last category (non-mechanical non-chemical) is a catch-all that includes everything that is not in one of the other two categories!

Elements Subject To Variation Should Be Described Tautologically

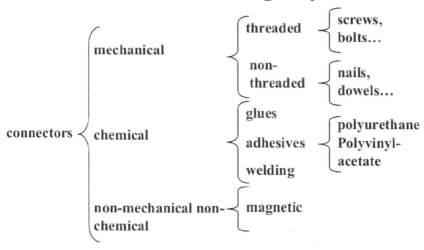

Figure 45

D) Use Overlapping Target Claiming

Patent attorneys almost always write claims that target the most preferred embodiment, and then expand out to claim less preferred alternatives. That strategy known as "target claiming" is fairly effective, but only if the patent attorney really understands where all the prior art lies. If someone comes along after the patent is issued, and uncovers some new prior art that anticipates or renders the broadest claims obvious, then the remaining claims may well be too narrow to keep the competition at bay.

The following example is a Venn diagram of four claims, with claim 1 being the broadest, and claims 2, 3, and 4 being successively narrower. The red box represents prior art that was located during trial. Since it falls within the scope of claims 1 and 2, those claims are subject to invalidation, leaving the inventor with only claims 2 and 4. In this example, the prior art destroys most of the value of the patent by invalidating (shown in lighter color) almost the entirety of claims 1 and 2.

Simple Target Claiming

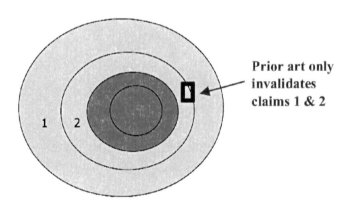

Prior art only invalidates claims 1 & 2

Figure 46

The best way to avoid this problem is to use overlapping target claiming. Here, the attorney focuses different sets of claims (preferably different sets of dependent claims) on different aspects of the invention. Thus, if one set of claims focuses on the materials employed in constructing a chair, another set of claims might focus on different coverings. As can be readily seen, the very same prior art that invalidated the entirety of claims 1 and 2 using ordinary target claiming, is only effective at destroying some of the coverage of those claims. Most of the rest of the space is protected by other claims.

Overlapping Target Claiming

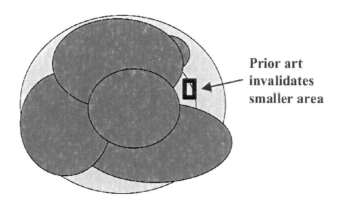

Prior art invalidates smaller area

Figure 47

E) Focus On Claims That Can Realistically Be Enforced

Cost-effective patenting not only focuses on *securing broad claims*, but it focuses on securing *claims that can be reasonably enforced* against an infringer. The two are not at all the same things.

Method of manufacture claims, for example, can be very difficult to enforce. Once a computer chip is manufactured, who can tell whether one of the polymer layers was cured at 100° F or 120°F? Enforcement of such claims requires discovery as to the exact manufacturing process of the defendant, and such discovery is exceedingly problematic unless opposing counsel is particularly generous. In the case of overseas manufacturing such discovery can be downright impossible to obtain.

If the invention does lie in the method of manufacture of the product rather than the end product itself, the patent attorney should work closely with the inventors to identify some difference, however small, between products produced by the inventive method versus other methods. Perhaps the novel manufacturing process produces material that is harder or softer than the previous methods, or is slightly smoother, or has some superior electrical property. In that case the claims should be directed to a product having the material with the claimed property.

Method of use claims can also be difficult to enforce. In the pharmaceutical field, for example, the generic drug manufacturers are not the ones administering the drugs. The manufacturers are merely providing drugs for physicians to administer, and can often escape infringement of method of use claims by asserting that they are selling the drugs solely for some non-infringing purpose. Indeed, the only way to establish infringement of a method of use claim may be to show that the manufacturer purposefully put his product into the marketplace to cause people to infringe the patent (inducement to infringe), or to establish that there are no substantial non-infringing uses of the product (contributory infringement). Those proofs can easily increase the cost of litigation by hundreds of thousands of dollars.

The excuses wear pretty thin when it comes time to justify method claiming in certain fields. As shown below, apparatus and method claims are often readily converted one into the other. But the apparatus claim is often enforceable against a manufacturer, while the method claim may only be enforceable against a consumer.

Apparatus Claims Can Be Easier To Enforce

Apparatus Claim

A bath soap dispenser, comprising:

a container sized and dimensioned such that the bath soap is applied to at least part of a body surface by movement of the container along the body surface;

wherein at least part of the bath soap is retained within the container during the movement; and

wherein the soap is advanced by pressure of a finger on the base plate.

Method Claim

A method of dispensing a bath soap onto a body surface, comprising:

providing a dispenser containing a base plate and a bath soap, wherein at least part of the soap is retained within the container during the movement;

advancing the soap by pressure of a finger on the base plate; and

moving the container along the body surface.

One should also focus on who or what entity can be sued. For example, where claims to methods of treating diseases in humans are unenforceable (as is the case in many foreign countries), it is better to claim a new chemical entity or a method of manufacturing than trying to claim the method of treating. Generally, doctors cannot be sued for treating patients even if the drugs they are using infringe upon a patent, but a manufacturer of the drug can be sued for producing and/or marketing an infringing drug.

Similarly, where an invention is a new use for an old machine, it is much better to claim (a) a controller, software or other new aspect that allows the old machine to perform its new function, and (b) to focus on the new purpose of the machine. But it is not at all uncommon to find an attorney who directly claims the new method. The attorney will likely get his claim issued, but may have a terrible time when he attempts to enforce it.

A third example is somewhat subtler. A composition claim is usually the Holy Grail in chemical applications. But in the field of polymers it is often the worst type of claim because polymer composition claims tend to be extremely narrow, and readily circumvented. From a litigation perspective it is much better to focus on claiming a class of polymers according to their characteristics, rather than claiming specific compositions.

More Enforceable	Less Enforceable
A computer chip comprising a dielectric layer having xyz properties....	A computer chip manufactured by laying down a first layer comprising ..., a second layer comprising ....
A method of shampooing hair using an anti-psoriasis compound...	A method of treating psoriasis using a shampoo having an anti-psoriasis compound....
A controller that cooperates with an xyz machine, using a protocol intended to treat a cancer in an organism.	A method of treating cancer, comprising using an xyz machine to irradiate the cancer ....
A method of imparting an image to a metallic surface comprising: ...coating the bonding area with a highly cross-linked polymer having a hardness greater than 80 and a coefficient of elasticity of at least 120% without breaking ...	A coating comprising 80–99.5 wt% of a polymerizable acrylate selected from mono-, di-, and triacrylates, urethane-modified acrylates, and polyester-modified acrylates, and 0.5 – 15 wt% of a photointiator...

F) <u>Avoid Means Plus Function Claims</u>

Another major category of difficult-to-enforce claims is means-plus-function claims. In the 1980s and early 1990 it was very fashionable to claim elements of an invention using the phrase "means for ..." in place of describing structural limitations. Thus, instead of identifying a doorknob as a "door knob", the patent attorney would instead describe the doorknob by its function, e.g. a "means for opening a door". That seemed very clever at the time because it appeared to be an easy way for patent attorneys to describe an invention without limiting the invention by specific structure. The problem, however, was that the absence of structure meant that the claims failed to satisfy their primary function of limiting the scope of the subject matter being claimed.

By the mid 1990s the Federal Circuit resolved that issue by limiting the scope of means-plus-function claims to the embodiments actually disclosed in the specification and drawing, and equivalents thereof. Thus, if a claim recites "means for opening a door" and the only such means disclosed in the application was a doorknob, the "means" would only include the disclosed doorknob and more or less exact equivalents, and would very likely exclude a latch type opener.

The scope of equivalents accorded a means-plus-function element is thus much narrower than the scope afforded other elements under the doctrine of equivalents. This is a complicated issue, and the reader is directed to the numerous judicial opinions on the subject. The important point here is that one should stay away from means-plus-function claims because they are excessively narrow.

Chapter VIII - File Foreign Only Where It Makes Sense

Foreign patent filings are expensive, many times more expensive that US filings. Foreign patents are also generally more expensive to enforce, and much less valuable when they can be enforced. Decisions on filing abroad should generally start with a budget, and then fit the foreign filing into the budget. Although the foreign countries in which an application should be filed differ from technology to technology, there is a general progression of countries in which foreign filings should occur. A caveat is that some types of inventions are not protectable abroad, or are only protectable using claim language that is even more twisted than normal.

The biggest financial problems in patenting typically come from the high costs of foreign patent prosecution. As a general rule of thumb, it costs about $4,000 - $5,000 per country or region to get a foreign patent application on file, and again that much per country by the time the patent issues. It gets even worse down the road, because most of the foreign countries have maintenance fees, about $600 per year. The short of it is that securing foreign patents will likely set you back about $10,000 per country, and about $25,000 per country over the life of the patent.

Given the high costs involved, it behooves an inventor/applicant to make some hard decisions about where he really wants to file. Some very large countries are probably not worth filing in because (a) their populations don't have a lot of money; (b) they don't have a reliable legal system; and (c) they have a cultural history of infringement. A particular country that comes to mind is Russia. Even if you obtain a patent in Russia, can you really enforce it?

The bottom line is that cost-effective patenting involves making hard decisions about where to file foreign applications. Obviously, the game plan is different for different technologies. In petroleum processing, for example, companies often file in Venezuela, Norway, and several of the Middle Eastern countries. In electronic chip manufacture, companies often choose to file in China and Taiwan. In general, however, we recommend the following filing schedule based upon the likely budget. Note that the dollar figures shown in the table reflect t he estimated costs of securing the patents, not just filing them. Thus, the costs for Western Europe are fairly high because one typically files national phase in seven or eight countries (Great Britain, France, Germany, Italy, Spain, and one or two others depending upon the product involved).

Select Foreign Filings According To Budget

USA	USA	USA	USA	USA
	W. Europe	W. Europe	W. Europe	W. Europe
		Japan	Japan	Japan
			China	China
			Canada	Canada
			Mexico	Mexico
				Korea
				India
				Brazil
				Taiwan
15,000	75,000	90,000	120,000	150,000

Figure 48

Not all types of inventions are patentable abroad. Methods of treating humans for diseases, for example, are often unpatentable in foreign countries. In some instances there are clever ways around that prohibition. Instead of claiming:

> "a method of treating psoriasis using a medicated shampoo having a <novel composition>"

it is often possible to claim:

"a method of shampooing hair by including <the novel composition> in the shampoo"

or

"a method of manufacturing a shampoo for use in treating psoriasis, by including <the novel composition> in the shampoo."

The language sounds the same to most people, but there is a huge difference to allowability of the claim.

Methods of doing business are also unpatentable in many foreign countries. Here again, there are often clever ways of circumventing the problem. For example, instead of claiming a method of marketing mortgages on the Internet, it is better to focus on claiming a computer system that runs software that markets mortgages on the Internet.

Basic information concerning foreign patents and foreign patent rights can be found at the website of the United States Patent Office (USPTO), at www.uspto.gov, specifically at www.uspto.gov/web offices/pac/doc/general/treaties.htm. Web sites for foreign patent offices are listed at www.uspto.gov/eb/menu/other.html. The World Intellectual Property Organization (WIPO) administers the Patent

Cooperation Treaty (PCT) applications, and hosts a website with a wealth of information regarding foreign patents. See www.wipo.int. PCT applications cover most the countries of interest for patent filings, with the notable exception of Middle Eastern countries. A good source of information for those countries is www.agip.com.

Prints of foreign patents can be difficult to obtain. Good sources are http://ep.espacenet.com (European patents), www.jpo.go.jp (Japanese patents), http://cipo.gc.ca (Canadian patents), and www.reedfax .com for all others.

Chapter IX - Make Sure Everyone Is Working Off The Same Page

A great deal of time and money is often wasted by allowing creativity to languish. In addition to merely drafting and filing patent applications, there are five specific areas in which your patent attorney should be actively facilitating the patenting process. Management can also greatly assist the patenting process by encouraging brainstorming and cooperation among inventors, by providing monetary and recognition incentive, and so forth. In addition, management should provide training in the "process of inventing" itself. There are basic principles underlying the evolution of all technologies, and those principles can be taught to inventors in a manner that vastly improves their inventing skills.

A) What Your Patent Attorneys Should Be Doing To Facilitate The Patenting Process

It is very frustrating to be in management when the people under you all pulling in different directions. The problem is exacerbated by the lack of experience that most people have in the field of patenting. This section is intended to give you suggestions on what your patent attorney should be doing.

What Your Patent Attorney Should Be Doing

- Work closely with your staff to view the invention in light of the prior art
- Help you focus on where you can create unique value
- Keep costs in line with the work performed
- Provide ongoing training
- Provide ongoing evaluation of the portfolio

Figure 49

- <u>Work closely with your staff to view the invention in light of the prior art</u>. Understanding the invention is the easy part of the task. The more difficult part is understanding how the invention fits into the evolution of the technology. This takes a rather unusual ability to view the invention from an appropriate level of abstraction.

- <u>Help you focus on where you can create unique value</u>. The inventor's job is to derive *one* good way of embodying the invention. Your patent attorney's job is to derive *all* of the commercially feasible variations and likely improvements. This is accomplished by reducing the inventive concepts down to their essence, and then figuring out how to patent that essence to be used as both a sword and a shield against the competition.

- <u>Keep costs in line with the work performed</u>. Cost-effective representation requires a mix of senior and junior patent attorneys (or patent agents), as well as paralegals, secretaries and other assistants. The work of those people needs to be allocated properly, with the most experienced people focusing mainly on brainstorming and claim drafting. Cost-effective representation also means that "standard" costs should be avoided in favor of "real" costs.

Just to cite one example, it is a common task to file a PCT application claiming priority to a utility application. The PCT filing is fairly simple, and involves little more than reformatting and reprinting the utility application, adding some filing pages, and placing the application in the mail. Many firms charge $1000 to $2500 for the office work involved (in addition to the PCT filing fees). You should be charged only for paralegal time and attorney review time, which probably comes to less than $250 in simple cases.

- Provide ongoing training. Inventors are usually well versed in their particular field, but know almost nothing about patenting. Many inventors, for example, erroneously think that an invention has to be made and tested before an application can be filed at the patent office. The reality is that a so-called "paper" patent is just as valid as any other patent. Many inventors also think that an inventive idea has to be developed to the stage where it works well before it can be patented. Wrong again. As long as the patent application explains the best way the inventor knows how to make and use the invention, the patent law is satisfied even if the best way known to the inventor works only poorly or intermittently. To derive the most benefit from their technical knowledge, inventors need to understand just what can be claimed. Your patent attorney is the person to do that training.

- <u>Provide ongoing evaluation of the portfolio</u>. Patenting is usually not a one-time process, at least not for companies. To make sure that you are receiving the most value for your money, your patent attorney needs to work with you in periodically reviewing the portfolio. Sometimes foreign matters need to be dropped. Sometimes entire families need to be abandoned. These are all things that should be discussed proactively with your attorney.

In short, your patent attorney is best used as an integral member of the development team. He should be helping you decide what to patent and how to patent it. He should be helping to train your people, and he should be proactively watching over costs, making sure that you get the most "bang for the buck."

B) <u>What Management Should Be Doing To Foster Creativity</u>

The best thing management can do to help the patenting process is to foster creativity. That can be accomplished in many ways, including hiring of creative people, providing surroundings that encourage brainstorming and cooperation, providing monetary and recognition incentive, and so forth. Most of these things are well known. But one thing that is often overlooked is the importance of management in helping the technology of their company to evolve. Management personnel have a great advantage in this respect, precisely because they are often not scientists, and are often <u>not</u> involved in the nitty gritty, day-to-day R&D efforts. That distance from daily involvement provides an opportunity to view the process from a perspective that others may lack.

Several years ago the Soviet government funded a project to identify how people invent. The project was never completed by the Soviets, but was eventually commercialized in the West as a sophisticated software package by a company named "The Invention Machine" (see www.invention-

<u>machine.com</u>.) One of the most interesting aspects of Invention Machine's software is its focus on general principles of evolution of technology. With those principles, it is possible to project how a technology will evolve, and make suggestions along those lines -- even if the person making the suggestions has only minimal knowledge of the technology.

I have further developed and simplified "evolution of invention" principles, and often apply those principles in brainstorming sessions with inventors. Here are some of the principles that I use:

Evolution Of Technologies

- Few to many
- Lower dimensions to higher dimensions
- Lower frequency to higher frequency
- Standalone system to feedback
- Stationary to mobile
- Generalized to specialized
- Single system to overlapping systems

Figure 50

These principles apply in an amazing number of cases. Consider for example, the process of cleaning teeth. The first device was undoubtedly a person's own finger or perhaps a stick, a single item moving a rate of about one hertz. The device has only a single element, operates at low frequency, has no feedback, has a generalized function in that the single finger handles all of the different types of tooth cleaning, and has no cooperation with other systems.

Without knowing anything else about cleaning teeth, one could use the Evolution of Technology principles to predict that the next development would be to go from few to many, and from lower dimensions to higher dimensions. In fact, the next big development was a brush, a device that has many "fingers" arranged in a two dimensional array. Can one go to a third dimension? Yes, by using bristles of different lengths.

Toothbrush Evolution
(Few to Many)
(Lower Dimensions To Higher Dimensions)

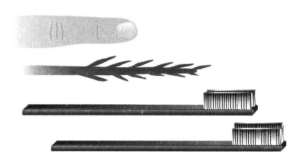

Figure 51

One could, of course, invent a toothbrush with ever more bristles, but there are only so many bristles that one can arrange in a small space, and one really can't go to a fourth dimension. So the next invention has to be going to a higher frequency. Since a hand operated toothbrush can only move at a few hertz, the next invention must be some sort of motorized toothbrush movement. We saw that, of course, in early electric toothbrushes that rotated back and forth at dozens or even hundreds of hertz.

That might have seemed like the end of the road to many inventors, because at higher and higher frequencies of movement the rotating toothbrush just falls apart. But using the invention principles it should have been obvious to ask how one could get to much higher frequencies. The answer, of course, was to use sonic frequencies. But then what? At ever-higher sonic frequencies the effects on tooth cleaning start to vanish. A clever inventor, however, would have predicted that in order to get even higher frequencies, one must start using light waves. Indeed, that is exactly what has happened. There are toothbrushes in which laser light emanates from the ends of the bristles, and lasers are also used to whiten teeth

Toothbrush Evolution
(Stationary to Mobile)
(Lower Frequency to Higher Frequency)

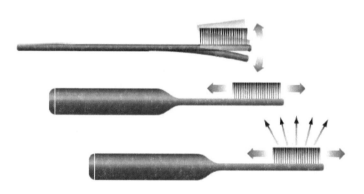

Figure 52

At some point the frequencies cannot realistically get any higher, so toothbrush development must start to involve some type of feedback. The newest toothbrushes change color when they need to be replaced. Toothbrushes of the future will have feedback in some other manner, such as interacting with a little display to keep track of when they were last used, and for how long. A toothbrush may even beep or play music when it hasn't been used in awhile, calling the errant child or adult to his tooth cleaning responsibilities!

What about the other principles? Toothbrushes are inherently mobile, and are readily carried about in a purse or suitcase. But the parts of a toothbrush should also become mobile. We have already seen toothbrushes having a handle that flexes in some manner. Toothbrushes in the future will have additional mobility.

Toothbrushes of the future will also become more and more specialized. We already have different toothbrushes for children, for adults, for those with sore gums, and so forth. Toothbrushes of the future will also have overlapping systems. There are already mechanical toothbrushes on the market that have a combination of rotating and reciprocating parts. Perhaps newer toothbrushes will combine rotating or reciprocating parts with sonic vibration, or perhaps diodes for producing laser light at a frequency that breaks up plaque. The point is that all of these concepts can and should be vetted during brainstorming sessions of engineers and other inventors.

A company may or may not have sufficient knowledge or resources to put together such sessions. But patent attorneys should be doing this work regularly, and they should be quite adept at drawing out new inventions from the inventors.

GLOSSARY

abandoned application

An application that is no longer pending. Applications can go abandoned because the applicant expressly abandons them, or because the applicant failed to respond to a final rejection. A parent application of often, but not always, abandoned when a child application is filed.

allowable claims

A claim that is deemed allowable by the patent office. Each claim of a patent application can be allowed or rejected independently of all other claims.

apparatus claim

A claim to a physical thing, such as a machine or a chemical composition. This contrasts with method claims, which are drawn to steps in a process.

application

A filing for a patent. An application can have a status of pending or abandoned. A formal patent application has an abstract, specification, at least one claim, and usually at least one page of drawing. The specification usually has a title and sections for field of the art, background, short description of the drawing, detailed description, and examples.

application filing

The date on which an application is filed. Filed applications are pending.

child application

An application that claims priority to one or more parent applications.

CIP application

A child application that contains additional disclosure relative to the parent. CIP applications have multiple priority dates, one to the filing date of the parent with respect to subject matter disclosed in the parent, and another to the filing date of the CIP with respect to the additional disclosure (termed "new matter").

GLOSSARY

claim drafting — The writing of patent claims, especially with an eye to broadly protecting a patentable invention.

claims — Numbered sentences following the patent specification, which define the scope of the claimed invention(s). Each claim covers a slightly different but overlapping scope.

co-inventor — An inventor that shares inventorship with another person. Intentional failure to list a co-inventor on a patent application may render any ensuing patent unenforceable.

commercially viable solution — An embodiment of an invention that is commercially significant. There are almost always many embodiments that are technically feasible, but commercially unimportant. One of the goals of patent drafting is to secure for the applicant patent rights to as many of the commercially viable embodiments as possible.

| continuation application | The term is strictly construed to mean a child application that supercedes the parent application. The USPTO used to refer to these continuations as FWC (file wrapper continuations) and used to issue a new serial number. The office then changed the name to RCE (Request for Continuing Application) and continued prosecution without changing the serial number. The latest incarnation is called a CPA (Continuing Patent Application), which also uses the same serial number as the parent, but now there is no pretense that the continuing application is anything other than a reincarnation of the parent. The term "continuing application" somewhat confusingly includes continuations, divisionals, and continuations-in-part. |

GLOSSARY

dependent
claims

A claim that is dependent on at least one other claim. The limitations of a dependent claim are those contained within the dependent claim, as well as all limitations contained within any claims upon which the dependent claim is directly or indirectly dependent. Thus, if claim 3 is dependent on claim 2, and claim 2 is dependent on claim 1, then claim 3 contains all the limitations of claims 1, 2, and 3.

disclosure

The subject matter described in a patent application, whether claimed or not.

divisional
application

A child application having the same specification as, and claiming priority to, a parent application. A divisional is usually employed to prosecute claims that were withdrawn or cancelled from the parent.

GLOSSARY

drafting
charges

Amount charged for writing the text of a patent application. The term is also sometimes used to mean costs associated with preparation of the drawing.

drawing

The figures of a patent. Technically there is only one drawing, even though the drawing may extend over several pages.

elements

Words or phrases of a patent claim that refer to a portion of the subject matter being claimed. Thus, in a claim to a chair, the elements may be the legs, seat, arms, back, coverings, connectors, etc.

enforceability

The ability to prevail against an infringer in a court of law on a claim of patent infringement.

examiner

The person at the patent office who reviews the prior art, and makes determinations as to

GLOSSARY

patentability. Examiners are not concerned with enforceability.

family   A group of at least two patents and/or patent applications that are linked by virtue of priority claims to one another. A patent family often has three or more "generations".

filing costs   Filings fees plus charges for completion and submission of the various papers that accompany a patent application.

filing date   The date that a patent application is considered to have been received by the patent office. The filing date is the same as the priority date if there is no priority claim.

filing fee   The fee charged by the patent office to accept a patent application for processing.

GLOSSARY

foreign
application

An application that is filed outside of the country having original filing. Thus, if a patent is originally filed in the United States and later in Japan, the Japanese application is a foreign application.

formal
application

An application other than a provisional application. This usually means a utility or PCT application. Formal applications must have at least one claim, whereas a provisional application need not have any claims.

grandchild
application

An application that claims priority to both a parent application, and a parent of the parent.

improvement

An embodiment of an invention that was not disclosed in a prior application.

independent
claims

A claim that is not dependent on any other claim. All of the limitations of the claim are

GLOSSARY

therefore contained within the independent claim.

informal application
A provisional application. Such applications are informal in that, among other things, they do not need to include any patent claims.

invention
Something that is new, useful, and non-obvious over the prior art.

invention-centered approach
A strategy that focuses on claiming an invention by its technical merits.

inventor
A person who conceived or helped conceive of an invention. A patent application can name multiple inventors. The head of a department, or other person who might well be listed on a journal article, is only an inventor for patent purposes if he/she actually contributed to the conception of the invention. Similarly, a person who helped build a prototype is

not necessarily an inventor, despite the fact that he/she may have contributed far more physical effort and time than an inventor. Inventors can be listed on a patent application in any order.

IP

Intellectual Property, which is generally considered to include patent, trademark, copyright, and trade secret rights.

large entity

In the United States, an assignee that has at least 500 employees. Many countries do not distinguish between large and small entities.

market-centered approach

A patenting strategy that focuses on claiming the commercially viable embodiments that preclude competition, rather than on the technical merits of the invention. Compare with invention-centered approach.

GLOSSARY

means-plus-function claims	A claim that includes at least one element that is defined by its function rather than a physical limitation (e.g., "means for opening a door" rather than "a door knob"). Means-plus-function claims do not necessarily have to include the term "means for".
method claims	A claim drawn to steps in a process rather than a physical thing *per se.* Method claims usually begin each phrase with a word ending in "ing", such as "enclosing", or "providing" or "connecting".
method of use claim	A type of method claim in which the applicant focuses on the manner in which something (often a Pharmaceutical or machine) is used.
multiple dependent claims	A claim that is alternatively dependent upon more than one claim. A typical format would be

GLOSSARY

"A device according to any of claims 1, 3, 4, or 7, in which ...."

office action

A formal communication from the patent office. Some office actions are favorable, some are unfavorable (rejections and objections), and some are informational only.

one year deadline

There are two one-year deadlines. A PCT application can only claim priority to an earlier filed application if the PCT application is filed within one year of the earlier filed application. Also, a provisional application will go abandoned unless a formal application is filed within one year of the provisional's filing date, and claims priority to the provisional.

patent application

An application to which one or more child applications claim priority.

GLOSSARY

patent attorney   An attorney who has passed the patent bar, and is responsible for drafting and/or prosecuting a patent application before a national or regional patent office. A patent agent is a non-attorney who has passed the patent bar, and can therefore draft and/or prosecute patent applications on behalf of others. Patent attorneys in one country can usually only practice law in that country.

patent drafter   The person or persons who draft the patent application. Even though the inventor may assist in the process, the task of correctly drafting a patent application ultimately falls to the responsible patent attorney or agent.

patent office   The national or regional authority charged with receiving and processing patent applications. In the United States the patent office is the USPTO.

GLOSSARY

PCT

Patent Cooperation Treaty; an international treaty signed by the United States, and administered by WIPO. The PCT receiving office for the United States is the United States Patent and Trademark Office (USPTO). Patent applications are examined through the PCT procedures, but the PCT never issues any patents.

petition to make special

A formal petition before the USPTO to speed up processing of a patent application based upon satisfaction of particular requirements.

preferred embodiment

A preferred implementation of the subject matter of a patent or patent application. Patent applicants in the United States are required to satisfy the "best mode" requirement, which means that they must describe whatever implementation of the claimed invention(s) that

GLOSSARY

they consider to be "best" at the time that the application is filed.

<table>
<tr>
<td>primary application</td>
<td>The oldest formal application in a family of patent applications. Subsequent (secondary) applications in the family usually focus on various subsets of the disclosure of the primary application.</td>
</tr>
<tr>
<td>prior art</td>
<td>Knowledge that is sufficiently close to the claimed subject matter that it is considered to be relevant to patentability. Prior art can be US or foreign patents, newspaper, journal or other publicly accessible documents, web pages, advertisements, and so forth. Prior art is defined by statute (35 USC § 102) for purposes of determining anticipation, but is not limited in that manner for purposes of determining obviousness.</td>
</tr>
<tr>
<td>priority; priority date</td>
<td>A legal fiction by which something that occurs later in time is treated</td>
</tr>
</table>

as if it occurred earlier in time. The claims of a divisional patent application, for example, have the filing date of the divisional application, but are treated as if that filing date were the filing date of the parent application for purposes of determining patentability.

provisional application	An informal patent application. Provisional applications are never examined. Unless they are used as a parent in a formal application, they are microfilmed and placed into storage at the on-year anniversary. In the latter case the provisional is then considered to be "dead" (expired).
reductionistic thinking	A process of reducing a complex idea, system, etc., to simpler parts or components that contain the essence of the idea or system.

GLOSSARY

rejected claims   Claims that the examiner considers to be unpatentable over the prior art, either because the claims are anticipated, obvious, and/or for some other reason. Claims that are merely objected to, rather than rejected, contain a technical defect that can usually be overcome relatively easily.

restriction requirement   A statement by the patent office that the pending claims address more than one invention. Restriction requirements are very commonly issued where an applicant has some claims directed to a method and some claims directed to an apparatus.

scope of equivalents   A patent claim covers both that which is literally encompassed by the language of the claim, and also that which is equivalent. The idea behind the doctrine of equivalents is that an infringer should not be able to circumvent a patent claim

GLOSSARY

by making an insubstantial modification.

small entity
: In the United States, an assignee that has less than 500 employees. Many countries do not distinguish between large and small entities.

target claiming
: A claiming strategy in which an independent claim recites a broad subject matter, and dependent claims recite successively narrower subsets of that subject matter.

tautological claiming
: A claiming strategy that logically includes all possible choices, even though the claim drafter does not know of all the choices.

USPTO
: United States Patent and Trademark Office

utility application
: A patent application that claims a useful invention. Contrasts with a design application, which claims the ornamental appearance of something.

GLOSSARY

Venn diagram   A diagram that uses circles and ovals to represent applications of set theory.

WIPO   World International Property Organization

INDEX

TABLE OF FIGURES